Sea-Level Changes
The Last 20 000 Years

Coastal Morphology and Research

Series Editor: Eric C.F. Bird

CORAL REEF GEOMORPHOLOGY
André Guilcher

COASTAL DUNES
Form and Process

Edited by
Karl Nordstrom, Norbert Psuty and Bill Carter

GEOMORPHOLOGY OF ROCKY COASTS
Tsuguo Sunamura

SUBMERGING COASTS
The Effects of a Rising Sea Level on Coastal Environments
Eric C.F. Bird

BEACH MANAGEMENT
Eric C.F. Bird

SEA-LEVEL CHANGES
The Last 20 000 Years
Paolo A. Pirazzoli

Sea-Level Changes

The Last 20 000 Years

Paolo Antonio Pirazzoli
CNRS, Meudon, France

JOHN WILEY & SONS
Chichester • New York • Brisbane • Toronto • Singapore

Copyright © 1996 by John Wiley & Sons Ltd,
Baffins Lane, Chichester,
West Sussex PO19 1UD, England

National   01243 779777
International (+44) 1243 779777

e-mail (for orders and customer service enquiries):
cs-books@wiley.co.uk
Visit our Home Page on http://www.wiley.co.uk
or http://www.wiley.com

Reprinted February 1998

Other Wiley Editorial Offices

John Wiley & Sons, Inc., 605 Third Avenue,
New York, NY 10158-0012, USA

Jacaranda Wiley Ltd, 33 Park Road, Milton,
Queensland 4064, Australia

John Wiley & Sons (Canada) Ltd, 22 Worcester Road,
Rexdale, Ontario M9W 1L1, Canada

John Wiley & Sons (Asia) Pte Ltd, 2 Clementi Loop #02-01,
Jin Xing Distripark, Singapore 129809

Library of Congress Cataloging-in-Publication Data

Pirazzoli, P. A. (Paolo A.)
    Sea level changes : the last 20 000 years / by Paolo Antonio
Pirazzoli.
        p.   cm. – (Coastal morphology and research)
    Includes bibliographical references and indexes.
    ISBN 0-471-96913-3
    1. Sea level.   I. Title.   II. Series.
GC89.P526   1996
551.4′58–dc20   96-25986
                                                          CIP

British Library Cataloguing in Publication Data

A catalogue record for this book is available from the British Library

ISBN 0-471-96913-3

Typeset in 10/12pt Times from author's disks by
Mayhew Typesetting, Rhayader, Powys
Printed and bound in Great Britain by
Biddles Ltd, Guildford, Surrey

This book is printed on acid-free paper responsibly manufactured from sustainable forestation, for which as least two trees are planted for each one used for paper production.

Contents

Preface

Studies of sea-level change have passed through several phases. It has long been realized that the sea has stood higher, relative to the land, at certain times in the geological past, and at other times lower, so that areas which are now sea floor emerged. A century ago these changes were ascribed to upward or downward movements of the land margin, the assumption being that sea level had remained more or less constant. Thus the term "raised beach" came into use to describe shore deposits that stood clearly above present sea level, the term implying that this was the outcome of tectonic uplift of the coastal fringe.

Just over half a century ago, opinion began to favour upward or downward movements of sea level – the so-called eustatic oscillations – as the primary explanation for emerged and submerged coastal features. The assumption was then that sea-level changes throughout the interconnected oceans had been equi-altitudinal, and that the sequence of emerged and submerged shorelines would eventually be correlated worldwide by their heights above and below present sea level. Implicit in this was the refinement of a "global sea-level curve", which would indicate the sequence of oscillations through Quaternary times. It was acknowledged that there would be discrepancies on coasts that were undergoing tectonic uplift, as in New Guinea, or subsidence, as in the Netherlands, where the equivalent shorelines would be found at higher or lower levels than those indicated on the global sea-level curve.

This was essentially how the subject was presented to me as a student in the University of London in the 1950s. During the past few decades, however, the concept of a global sea-level curve, applicable to tectonically stable coasts, has been abandoned because detailed studies on many coasts have found that sea level has not shown worldwide and equivalent oscillations, while continents drifting horizontally have had variable uplift or depression along coastal margins. Studies of sea-level changes have focused on local sequences, and contrasted histories have been established for various coastal regions, initially in relative terms, for it has proved difficult to separate land movements from sea-level oscillations.

Anyone who attempts a global assessment of sea-level changes must have experience from a variety of coastal regions and a grasp of the prolific and dispersed literature on this subject. Paolo Pirazzoli has these credentials. His research into sea-level changes began appropriately in his native Venice, a notoriously subsiding region, in the 1970s, and then extended round the Mediterranean and to the coasts of the Indian, Atlantic and Pacific Oceans. As Director of the International Geological Correlation Programme Project 200 ("Sea-level Correlation and Applications") from 1983 to 1987 he organized a review which, as well as generating many scientific reports, led to the publication of his benchmark *World Atlas of Holocene Sea-Level Changes* in 1991. The present work extends the time scale from the Holocene (10 000 years) to 20 000, in order to encompass the changes that have occurred since the last glacial phase of the Pleistocene, when global climate became exceptionally cold, and the world's oceans fell to very low levels. Subsequently there has been climatic amelioration, and a major marine transgression (known variously as the Late Quaternary, the End Pleistocene–Early Holocene, or the Flandrian transgression), in the course of which the major features of the world's coastlines have taken shape.

As Series Editor I invited Paolo Pirazzoli to address the problem of establishing sea-level change sequences during this period, and put them in a global context. This he has done with his usual energy and enthusiasm. I expect the review will prove to be timely as a means of stimulating and guiding discussion, and of preparing the way for the next phase of research into dated sequences of features indicating past sea levels, particularly on coasts that have so far received little attention. I hope that researchers in many countries will take on this task, and thereby improve our understanding of the intricate environmental changes that have taken place in the last 20 millenia.

Eric Bird
Oxford, April 1996

Acknowledgements

This book is a contribution to the activities of the IGCP Project 367 "Late Quaternary coastal records of rapid change" and of the INQUA Commissions on Quaternary Shorelines and on Neotectonics.

The author is grateful to the Series Editor, Eric C.F. Bird, for assistance in revising the English text, and to Mrs A. Sevestre (CNRS, Meudon) for the preparation of Figs. 65–69. For other illustrations, the author is grateful for permission to reproduce the following copyright material. Fig. 1: reprinted with permission from *Nature*, vol. 345, pp. 708–710, Eustatic sea level fluctuations induced by polar wander, R. Sabadini, C. Doglioni and D.A.Yuen, © 1990 Macmillan Magazines Limited. Figs. 2, 3 and 114: reprinted from *Global and Planetary Change*, vol. 8, no. 3, Global sea-level changes and their measurement, P.A. Pirazzoli, © 1993, with kind permission from Elsevier Science – NL, Sara Burgerhartstraat 25, 1055 KV Amsterdam, The Netherlands. Fig. 5: reprinted with permission from *Geological Society of America Bulletin*, vol. 78, pp. 1477–1494, A.L. Bloom, 1967. Fig. 6: reprinted with permission from *Quaternary Research*, vol. 11, pp. 279–298, J.A. Clark and C.S. Lingle, © 1979 University of Washington. Figs. 10 and 112 (B.C. Douglas, R.E. Cheney and R.W. Agreen, *Journal of Geophysical Research*, vol. 88 (C14), pp. 9595–9603, 1983) and Fig. 111 (R.E. Cheney, J.G. Marsh and B.D. Beckley, *Journal of Geophysical Research*, vol. 88 (C7), pp. 4343–4354, 1983), copyright by the American Geophysical Union. Fig. 12: reprinted from G. Russell in *Intertidal and Littoral Ecosystems*, pp. 43–65, ©1991, with kind permission from Elsevier Science. Figs. 27 and 36: reprinted with permission from *Journal of Coastal Research*, vol. 10, pp. 395–415, J. Laborel and F. Laborel-Deguen, © 1994 Coastal Education and Research Foundation (CERF). Figs. 29, 108, 109 and 110: reprinted with permission from CERF. Fig. 34: reprinted from *Marine Geology*, vol. 120, pp. 203–223, Biological evidence of sea-level rise during the last 4500 years, on the rocky coasts of continental southwestern France and Corsica, J. Laborel, C. Morhange, R. Lafont, J. Le Campion, F. Laborel-Deguen and S. Sartoretto, © 1994, with kind permission from Elsevier Science. Fig. 50a,b:

reprinted with permission from *Proceedings of the Geologists' Association*, vol. 93, p. 60, I. Shennan. Fig. 51: reprinted with permission from *Comptes Rendus de l'Académie des Sciences*, vol. 319, II, pp. 65–77, Climate instabilities: Greenland and Antarctic records, J. Jouzel, C. Lorius, S. Johnsen and P. Grootes, Editions Gauthier-Villars. Figs. 53, 54 and 56: reprinted with permission from *Science*, vol. 265, pp. 195–201, Ice age paleotopography, W.R. Peltier, © 1994 American Association for the Advancement of Science. Fig. 57: reprinted with permission from *Nature*, vol. 276, pp. 680–683, A glacial Mediterranean, J. Thiede, © 1990 Macmillan Magazines Limited. Fig. 58: reprinted with permission from *Journal of Coastal Research, Special Issue*, no. 1, pp. 95–98, P.A. Kaplin and F.A. Shcherbakov, © 1986 CERF. Fig. 60: reproduced from *Episodes*, vol. 9, no. 1, March 1986. Fig. 61: reprinted with permission from *Science*, vol. 241, pp. 440–442, The position of the Gulf Stream during Quaternary glaciations, T. Keffer, D.G. Martinson and B.H. Corliss, © 1994 American Association for the Advancement of Science. Fig. 63: reprinted from *Global and Planetary Change*, vol. 7 (1–3), Introduction to Quaternary earth system changes, H. Faure, L. Faure-Denard and T. Liu, © 1993, with kind permission from Elsevier Science. Fig. 64: reprinted with permission from *Quaternary Research*, vol. 16, pp. 125–134, W.F. Ruddiman and A. McIntyre, © 1979 University of Washington. Figs. 71, 74, 75 and 78: reproduced from P.A. Pirazzoli, *World Atlas of Holocene Sea-Level Changes*, © 1991, with kind permission from Elsevier Science. Fig. 72: reproduced from L.A. Dredge and F.M. Nixon, *Glacial and Environmental Geology of Northeastern Manitoba*, Memoir 432, 1992, by courtesy of the Geological Survey of Canada. Fig. 73: reproduced from *Zeitschrift für Geomorphologie*, vol. 24, C. Hillaire-Marcel and S. Occhietti, © 1980 Gebrüder Borntraeger. Fig. 77: reproduced from *Marine Geology*, vol. 124, pp. 1–19, Holocene evolution of relative sea level and local mean high water spring tides in Belgium – a first assessment, L. Denys and C. Baeteman, © 1995, with kind permission from Elsevier Science. Figs. 86 and 94: reproduced with permission from *Zeitschrift für Geomorphologie, Supplement* 102, pp. 21–35, P.A. Pirazzoli, J. Laborel and S.C. Stiros, © 1996 Gebrüder Borntraeger. Figs. 89 and 90: reproduced from *Journal of Geophysical Research*, vol. 101 (B3), pp. 6083–6097, P.A. Pirazzoli, J. Laborel and S.C. Stiros, © 1996 by the American Geophysical Union. Fig. 92 and 121: reproduced with permission from *Zeitschrift für Geomorphologie, Supplement* 40, pp. 127–149, Y. Thommeret, J. Laborel, L.F. Montaggioni and P.A. Pirazzoli, © 1981 Gebrüder Borntraeger. Fig. 97: reproduced from *Marine Geology*, (1996, in press), Holocene sea-level rise and shoreline positions in the southern North Sea, D.J. Beets, K.F. Rijsdijk, C. Laban and P. Cleveringa, with kind permission from Elsevier Science. Fig. 103: reproduced from *Reviews of Geophysics*, vol. 26, pp. 624–657, A. Berger, © 1988 by the American Geophysical Union. Fig.

104: reproduced from G.A. Maul, M.D. Hendry and P.A. Pirazzoli, Sea level, tide and tsunamis, in G.A. Maul (ed.) *Small Islands: Marine Science and Sustainable Development*, © 1996 American Geophysical Union (in press). Figs. 105, 106 and 107: reprinted from D.B. Scott, P.A. Pirazzoli and C.A. Honig (eds.) *Late Quaternery Sea-Level Correlation and Applications*, NATO ASI Series C, vol. 256, 1989, pp. 153–167, with permission of Kluwer Academic Publishers. Fig. 113: reprinted with permission from *Nature*, vol. 369, pp. 48–51, Rising temperatures in subtropical North Atlantic Ocean over the past 35 years, G. Parrilla, A. Lavi, H. Bryden, M. Garcia and R. Millard, © 1990 Macmillan Magazines Limited. Fig. 115: reprinted with permission from *Journal of Coastal Research*, vol. 9, pp. 104–111, Y.I. Ignatov, P.A. Kaplin, S.A. Lukyanova and G.D. Solovieva, © 1993 CERF. Fig. 118: reprinted from *Marine Geology*, vol. 110 (1), Marine deposits of late glacial times exposed by tectonic uplift on the east coast of Taiwan, P.A. Pirazzoli, M. Arnold, P. Giresse, M.L. Hsieh and P.M. Liew, © 1993, with kind permission from Elsevier Science. Fig. 119: reprinted with permission from *Nature*, vol. 382, pp. 241–244, Deglacial sea level record from Tahiti corals and the timing of global meltwater discharge, E. Bard, B. Hamelin, M. Arnold, L. Montaggioni, G. Cabioch, G. Faure and F. Rougerie, © 1996 Macmillan Magazines Limited. Fig. 120: reproduced with permission from *Zeitschrift für Geomorphologie, Supplement*, vol. 62, pp. 31–49, P.A. Pirazzoli, © 1996 Gebrüder Borntraeger. Fig. 122: reprinted from *Global and Planetary Change*, vol. 8(3), Estimations of a global sea level trend: limitations from the structure of the PSMSL global sea level data set, M. Gröger and H.P. Plag, © 1993, with kind permission from Elsevier Science.

The following authors are thanked for permission to reproduce figures: C. Baeteman, E. Bard, A. Berger, E.C.F. Bird, A.L. Bloom, M. Bondesan, D.J. Beets, S. Carboni, R.E. Cheney, L.A. Dredge, H. Faure, D.R. Grant, J. Jouzel, P.A. Kaplin, J. Laborel, D.G. Martinson, G.A. Maul, N.A. Mörner, L.F. Montaggioni, S. Occhietti, G. Parrilla, R.W. Peltier, H.P. Plag, W.F. Ruddiman, G. Russell, I. Shennan, D.E. Smith and J. Thiede. Special thanks go to C. Baeteman, E. Bard, E.C.F. Bird, R.E. Cheney, H. Faure, J. Laborel and G. Parrilla for providing good quality reproductions of illustrations.

Unless otherwise stated, all the photographs were taken by the author.

Introduction

Sea-level changes have interested and puzzled several generations of scientists. Significant vertical displacements of land in relation to average mean sea levels are generally rare over the time scale of a human life. This led most observers to believe that the land level was stable. Divergent observations on local sea-level changes in the past have therefore been difficult to interpret and explain. In addition, the strong influence of certain inherited ideas did not help to clarify the situation until very recently.

If the ocean surface could not slope up or down, but had to remain at the same level above the centre of the Earth, as believed by the Greek geographer Strabo and confirmed 13 centuries later by Dante (Allighieri 1320), then sea level should be the same all around the globe. Evidence of emergence or submergence along certain coasts was explained in the past in a number of ways: as a result of gradual uplift and tilting on the Baltic coasts, of irregular uplift on the coast of Chile (Darwin 1846), of subsidence around the Mediterranean (Manfredi 1746), or of changes in the ocean water volume (Maillet 1748). A detailed discussion on ancient theories explaining sea-level changes was published by Suess (1885).

At the end of the 17th century, and especially in the 18th century, the first scientific observations for the purpose of measuring sea-level changes were made. This was attempted, after the establishment of the first tide-gauge datum in 1682 at Amsterdam (see Section 6.2.1), by a physicist, Hyarne, who in 1702 cut some marks on rocks in Sweden, and by B. Zendrini, who in 1732 used a step of the Doge's Palace as a sea-level datum in Venice (Zendrini 1802).

Suess (1885) introduced the concept of "eustatic" changes in sea level, i.e. vertical displacements of the ocean surface occurring uniformly throughout the world. Daly (1934) stressed the importance of changes in sea level and of "glacio-isostatic" effects accompanying the last deglaciation phase, with uplift in areas of ice melting and subsidence in a wide peripheral belt (see Section 1.3).

Studies of Late Quaternary sea-level changes have greatly improved during the last few decades owing to the existence of four favourable conditions:

(1) the development of relatively precise radiometric dating methods (e.g. radiocarbon, U/Th); with some procedures (e.g. accelerator mass spectrometry and thermal ionization mass spectrometry), very small samples are now sufficient to obtain a relatively accurate date;
(2) the activity of successful international research programmes (e.g. International Geological Correlation Programme, IGCP, and International Quaternary Association, INQUA) which have favoured an interdisciplinary approach, attracting contributions from geologists, geomorphologists, geographers, oceanographers, geophysicists, archaeologists, marine biologists and glaciologists, and have made possible a wide geographical coverage of field work;
(3) the improvement and increasingly common use of powerful computers, which have made possible the development of complex isostatic models, and a comparison of field data with model results;
(4) the advent of satellite sensing, which has revealed that, contrary to Strabo's and Dante's belief, variations exist on the surface of the oceans, with a relief of as much as 200 m.

A more complete approach to sea-level changes, made possible by these favourable conditions, has brought about a revolution in theories and ideas. Many new publications on deglacial and postglacial sea-level changes have appeared during the last two decades, as a result of the growing scientific interest of researchers from several disciplines on this theme.

A debate started in the late 1950s and early 1960s, when divergent interpretations were proposed by Godwin et al. (1958), Fairbridge (1961), Jelgersma (1961), Shepard (1963), and several others, to summarize the shape (gradually rising, or fluctuating) of the curve that was assumed to represent the global eustatic Holocene sea-level rise.

Since 1974, UNESCO and the International Union of Geological Sciences (IUGS) have been sponsoring, in the framework of the IGCP, a series of research projects related to sea level: Project 61 "Sea-level changes during the last hemicycle" (directed by A.L. Bloom from 1974 to 1982), Project 200 "Sea-level correlation and applications" (directed by P.A. Pirazzoli from 1983 to 1987), Project 274 "Coastal evolution in the late Quaternary" (directed by O. van de Plassche from 1988 to 1993), and Project 367 "Late Quaternary coastal records of rapid change" (directed by D.B. Scott since 1994). Each of these projects attracted several hundred participants from 50 to over 70 countries, organized international and regional meetings on all continents and produced books and numerous scientific publications.

During the same period, two INQUA bodies, the Commission on Quaternary Shorelines and the Neotectonic Commission, were active in this field, also organizing meetings and publishing results, often in close collaboration with the contemporary IGCP Project.

The main aim of this wide international collaboration was a better understanding of the nature, causes, and impacts of sea-level changes during the late Quaternary. After two decades of collaborative work the main interests of the sea-level community are gradually shifting from the understanding of global and local sea-level changes *per se*, to a wide area of related fields and applications, such as the study of the geomorphological and sedimentary impacts of sea-level change, the rheological properties of the Earth which can be deduced from isostatic responses to ice or water loads on its surface, the identification of past coastal hazards (earthquakes, tsunamis, storm surges) from the stratigraphical record or from other remains, the mapping of coastal areas at risk to flooding, the production of geographical information systems and other data bases, and several other useful applications.

Following a period of unprecedented research activity and advancement in the understanding of how and why sea levels have changed, and preceding a possible swarming of part of the sea-level community towards other related tasks, it seemed timely to attempt to review knowledge of global and local sea-level changes since the last glacial maximum.

The present work differs from previous publications on similar topics, mainly because it is the first recent book on sea-level changes in which all the chapters are written by the same person. Its approach may therefore be less wide thematically than that of a multi-authored book, but more general and homogeneous, and not limited to a range of particular aspects or regions. It complements the *World Atlas of Holocene Sea-Level Changes* (Pirazzoli 1991), but differs from that work in approach, content and resolution. It also extends the time range considered to the last 20 000 years and addresses a less specialized readership.

Sea-level changes since the last glacial maximum have varied greatly from place to place. Formerly glaciated areas, when unloaded of ice, have generally shown a relative sea-level fall, often of the order of hundreds of metres. Oceanic waters, which received meltwater from ice sheets, have been deepened by the glacio-eustatic rise. The ocean floor has been depressed by the meltwater load. Coastal areas have reacted to the sea-level rise in a very variable manner, depending on their distance from former ice sheets, local topography and water depth. The result has been a wide range of regional relative sea-level histories, even in areas devoid of active crustal movements. In this book, after a discussion of the main causes of sea-level change (Chapter 1), of the methods employed to recognize and date former shorelines (Chapter 2), and an account of the ice age Earth (Chapter 3), examples are given of significant sea-level histories and tendencies around the world (Chapters 4 and 5). Present-day sea-level trends, their methods of measurement and possible relationships with an increasing greenhouse effect and other human influences are also discussed (Chapter 6).

It is shown that relative sea-level changes can be used to assess trends of tectonic movements, to recognize and date seismic displacements, to infer climatic changes and to improve geophysical models of the Earth's interior.

As the study of sea-level changes concerns several disciplines (geography, geology, oceanography, geophysics, prehistory, coastal archaeology, marine biology and ecology), this book addresses mainly an audience of post-graduate students and young researchers in these disciplines. It may also be useful to persons teaching various branches of Earth sciences who wish to update their ideas, to undergraduate students in geosciences, as well as to coastal ecologists and people interested in various aspects of global change. Applications of sea-level changes may also be of interest to coastal engineers, coastal zone managers and land planners.

Chapter One

Causes of sea-level change

1.1 INTRODUCTION

The sea surface is very unstable. It rises and falls with tide, waves, changes in the atmospheric pressure, wind, temperature and salinity. However, when periodical and random movements are filtered out, a stable value can be obtained: mean sea level (MSL). Altimetric measurements in most countries are related to a reference level which corresponds to MSL calculated at a selected tide-gauge station over a specified period.

In reality, accurate levellings have shown that along the coasts of a continent MSL varies from place to place, and at each place it also varies over time (see also Section 6.2.3.1). In Britain, for example, the Ordnance Survey refers all levels to Ordnance Datum Newlyn (ODN) which is the MSL determined from six years of continuous tide-gauge records at Newlyn, Cornwall, extending from May 1915 to April 1921. MSL at Newlyn is now about 0.2 m above Ordnance Datum (Hinton 1992). On the other side of the Channel, the zero datum of the Nivellement Général de la France (NGF) was determined at Marseilles from tide-gauge records of the period 1885–1897. Today MSL at the Marseilles tide-gauge station is about 11 cm above the NGF zero. It is not surprising, therefore, that to find the elevation at which the British and French national levelling systems could fit together in the Channel Tunnel was no easy task.

As a generalization, if tidal forcing, differences in water density, currents and atmospheric forcing are left aside, MSL can be defined as an equipotential surface of the gravity field. It will be modified, therefore, by any factor affecting the gravity force from outside the Earth (astronomical factors), at the Earth's surface (volcanic, ice-sheet or sedimentary loads), or in the Earth's interior (displacements of deep materials with different densities). In addition, any mass redistribution affecting gravity will change the moment of inertia of the Earth. Also involved will be the angular velocity and the polar drift. Acceleration of the angular velocity would

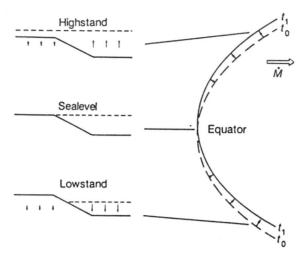

Figure 1   Pictorial representation of sea-level fluctuation induced by polar drift. The white arrow denotes the direction of polar drift. The solid and dashed curves represent the deformation of geoid and topography of the sea floor at times $t = t_0$, when the perturbation is initially imposed, and a subsequent time $t = t_1$. Sea level is unperturbed at the equator but rises and falls occur in the northern and southern hemispheres, respectively (after Sabadini et al. 1990)

cause a rise of sea level in the equatorial belt and a lowering at the poles, with changes of essentially the same order for points of equal latitude (Fairbridge 1961). Alternatively, a change in axis (polar drift) would produce deformation of the geoid, with apparent sea-level stability at the equator, but opposite movements of rise and fall in the two hemispheres, depending on the direction of polar drift (Sabadini et al. 1990) (Fig. 1).

When phenomena such as variations in angular velocity or polar drift are not taken into account, relative sea-level changes at any coastal site can be considered to result from the superimposition of two completely independent movements: that of the sea surface and that of the land surface. The sea surface can be modified by changes in the quantity of oceanic water, deformation of the shape of the oceanic basin, variations in water density, and dynamic changes affecting the water masses.

1.2   CHANGES IN THE QUANTITY OF OCEANIC WATER

The quantity of water in the oceans depends on the global hydrological balance. The equation for the water balance for the whole globe can be presented as follows:

$$A + O + L + R + M + B + S + U + I = K$$

Table 1   Estimates of storage volumes and average residence times of parameters of the world water balance

	Parameter	Present volume $(km^3)^{(a)}$	Equivalent water depth[b]	Average resident time[a]
A	Atmospheric water	13 000	36 mm	8–10 days
O	Oceans and seas	1370×10^6	3.8 km	4000+ years
L	Lakes and reservoirs	125 000	35 cm	often controlled by man
R	Rivers, channels	1700	5 mm	2 weeks
S	Swamps	3600	10 mm	of the order of years
B	Biological water	700	2 mm	1 week
M	Moisture in soils and the unsaturated zone	65 000	18 cm	2 weeks to 1 year
U	Ground water	4×10^6 to 60×10^6	11 to 166 m	From days to tens of thousands of years
I	Frozen water	32.5×10^6	90 m	tens of thousands of years

(a) Data from UNESCO (1971) except for I (from Oerlemans 1993)
(b) Using present ocean surface area $(361.3 \times 10^6 \ km^2)$

in which letters correspond to approximate water quantities summarized in Table 1, K being a constant value. Atmospheric water, rivers and channels, marshes, peat bogs, biological water and soil moisture represent altogether less than 24 cm of sea-level equivalent and can therefore be neglected for the moment. Lakes and reservoirs have been retaining increasing quantities of water during the last decades, but their impacts on sea level are mainly seasonal.

The volume of underground water is very poorly known and has probably changed considerably during the last 20 ka (ka = thousand years). The Sahara Desert, for example, was much wetter during the early Holocene, when lakes existed in areas that are now very dry (Petit-Maire 1986). Here, phreatic levels have dropped by between 10 and 100 m, depending on the area considered, during the second half of the Holocene. In many geological basins, existing underground sheets of water are several thousand years old (i.e. fossil), as shown by isotopic and geochemical analyses; excessive pumping, during the last century, has produced not only sediment compaction and land subsidence, but has also transferred water into the ocean basins and so contributed to a fraction of the recent sea-level rise recorded by tide gauges (Section 6.2.1).

The development or melting of continental ice sheets is of paramount importance for sea-level changes during the last 20 ka. The present ice volumes are known with fair accuracy (Table 2).

Climate change is indeed the main cause of changes in the quantity of oceanic water. According to oxygen isotope records obtained by analysing foraminifera in deep-sea sediment cores, initiation of moderate-sized ice

Table 2 Volume and equivalent ocean water depth of the main ice-caps, at glacial and present time

Glacier	Glacial time			Present time		
	Ice volume[a] (10^6 km^3)	Equivalent water depth[b] (m)	%	Ice volume[a] (10^6 km^3)	Equivalent water depth[b] (m)	%
Antarctica	37.7	104.3	40–56	27.9–29.3	77.2–81.1	90–91
Greenland	2.9–5.6	8.0–15.5	4–6	2.5–3.0	6.9–8.3	8–9
North America	18.0–36.7	49.8–101.6	27–39			
Eurasia	8.2–14.3	22.7–39.6	12–15			
Others				0.2	0.6	1
TOTAL	66.8–94.3	184.8–261.0	100	30.6–32.5	84.7–90.0	100

(a) Data from Hughes et al. (1981), Fisher et al. (1985), Berger et al. (1990) and Oerlemans (1993)
(b) Without hydro-isostatic sea-floor displacement, using present ocean surface area

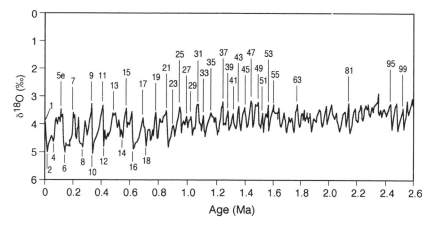

Figure 2   Oxygen isotope record for the past 2.6 Ma deduced from benthonic foraminifera of the ODP core 677 (after Shackleton et al. 1990). Labels of selected isotope stages have been added for orientation. Oscillations involve differences in ocean level of the order of 100 m recurring every 100 ka on average, with superimposed minor oscillations of 41 ka and 23 ka. A global sea-level rise of about 100 m is estimated to have occurred during the last 20 000 years

sheets in the northern hemisphere occurred about 2.40 Ma ago (Ma = million years), possibly preceded by small increases in the Earth's ice volume about 2.55 Ma ago and by a longer-term high-latitude cooling that began about 0.75 Ma earlier (Ruddiman and Raymo 1988). After 0.9 Ma, changes in $\delta^{18}O$ (oxygen isotope ratio) increased in amplitude by about 50%, suggesting that ice sheets grew to considerably larger volumes. Climatic oscillations during the past 0.45 Ma show a predominance of the 100 ka period of orbital eccentricity, with numerous superimposed colder phases (glacial advances) at periods of 41 ka (orbital obliquity) and 23 ka (precession cycle) (Fig. 2).

   During Stages 12 and 16, which correspond to exceptionally large glaciations, the ice volume was about 15% greater than at the last glacial maximum (Stage 2). Stage 6 was perhaps marginally more extreme than Stage 2 and Stage 10 marginally less extreme, while Stages 4, 8, 14 and 18 were significantly less important (Shackleton 1987). On the other hand, the warmest interglacials (with a higher sea level) occurred at Stages 1, 5e, 9 and 11, which are all so similar, according to Shackleton (1987), that it cannot be stated confidently that sea level attained in any one of these interglacials differed significantly from any other.

   Global sea-level changes during the last 250 ka, according to various estimates, are represented in more detail in Fig. 3; curve A was obtained from correlation between the coral terraces of Huon Peninsula and the ^{18}O record from core V19-30; curve B was estimated using planktonic and

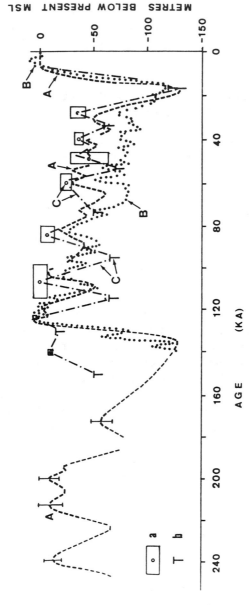

Figure 3 Eustatic sea-level curves for the past 250 ka, according to (A) Chappell and Shackleton (1986), (B) Shackleton (1987) and (C) Bloom and Yonekura (1990). Boxes indicate probable errors of age and height of sea-level maxima (a) and minima (b) related to curve C. Vertical bars are uncertainty limits of altitude related to curve A (from Pirazzoli 1993)

benthonic data from various oceanic cores; and curve C is an estimate of
the interstadial high sea levels of the last 105 ka deduced, with assumptions, from Huon Peninsula terraces. The possible causes of the difference
between curves A and B during the last glacial stage are still debated.
According to the latest results available (Chappell et al. 1994), curve B has
been confirmed by several new U/Th datings and is, therefore, more
reliable.

1.3 DEFORMATION OF THE SHAPE OF
THE OCEANIC BASIN

The deformation of the Earth's crust by tectonic movements produces
relative sea-level changes that differ from place to place. The main cause of
tectonic movements is probably isostatic imbalances.

1.3.1 Isostatic changes

The word "isostasy" (from Greek: *isos* = equal and *stasis* = equilibrium)
expresses the theory that all large portions of the Earth's crust are in
balance as though they were floating on a denser underlying layer; thus
areas of less dense crustal material tend to rise topographically above areas
of more dense material.

Satellite sensing has revealed that the sea-surface topography does not
correspond exactly with that of a perfect ellipsoid, but that there are bumps
and depressions, with a relief of up to 200 m. These irregularities have been
ascribed to density changes inside the Earth and therefore to isostatic
adjustments at the Earth's surface. As the Earth's interior is mobile,
displacements of materials with different densities can take place, changing
the gravity pattern of the Earth. Sea water, being liquid, will adapt instan-
taneously to any gravity change. The lithosphere will also adjust but, being
denser than water and more rigid, with less deformation occurring over
a much longer period. The resulting time lag and difference in vertical
shifting between the isostatic adjustments of water and land will cause
relative sea-level changes at a regional or even a continental scale. Very
little is known of the rates and durations of displacement of geoidal bumps
and depressions in the past, or how they may change in the future.

Plate tectonics slowly but continuously modifies the shape and volume of
the oceanic basins, and therefore global sea level. At a local scale, as the
ocean crust spreads away from its area of origin along submarine ridges, it
cools and thickens, thereby increasing in density so that the sea floor subsides isostatically (*thermo-isostasy*), and oceanic islands become gradually
submerged as they migrate. In tropical waters the subsidence is increased by
the load of coral reefs which grow upward and so thicken as they maintain
their shallow-water position. Average subsidence rates have been estimated

at 0.2 mm/yr over the past 60 Ma in the Marshall Islands (Menard and Ladd 1963) and 0.12 mm/yr since the Pliocene in Mururoa Atoll (Labeyrie et al. 1969).

Where the migrating oceanic crust approaches a hot spot (a lava source in the mantle), the normal cooling process is reversed and the crust is heated, becoming less dense and thinner, and so rises isostatically. This is the case, for example, of Anaa Atoll in the Tuamotu Islands, which is presently located about 200 km east of the hot spot area of the Society Islands and is approaching it, following the movement of the oceanic plate, at the rate of 45 mm/yr. Although the uplift rate is rather slow (0.1 mm/yr), it can last for a few million years, creating uplifted atoll features (Pirazzoli et al. 1988a; Pirazzoli 1995). However, moving away from a hot spot, there is renewed cooling and isostatic subsidence (Menard 1973; Detrick and Crough 1978).

In plate convergence areas, sea-floor sediments capping the underthrust oceanic plate are often too light to be subducted; they will pile up, forming a so-called accretion prism near the edge of the overthrusting plate, which will consequently be raised isostatically. On passive continental margins, the velocity of subsidence becomes very low: a value of 0.03 mm/yr over the last 135 Ma, which includes not only lithospheric cooling but also the effects of sediment loading, has been suggested by Gornitz (1993) for the east coast of the United States.

Isostatic deformation can also be caused by substantial loads charging the lithosphere surface. This will produce an isostatic depression under the load, and a slightly raised marginal rim at some distance from the load barycentre. Such a deformation can result from the load of an ice sheet, of a volcano, or of any major accumulation of sediments, or even of water.

As shown by Daly (1934), the development of an ice sheet will result in subsidence beneath the ice mass (*glacio-isostasy*), when deeper material flows away and a peripheral bulge is built around the ice margin. When the ice sheet melts, unloading occurs, resulting in uplift beneath the melted ice at rates which may locally reach values of the order 50 to 100 mm/yr (see Figs. 72, 73); the marginal rim will consequently tend to subside and move towards the centre of the vanishing load (Fig. 4).

Repeated extrusion of lava will generate isostatic processes (*volcano-isostasy*) comparable with those of an ice sheet of equivalent mass (Hamilton 1957; Walcott 1970a). Empirical calculations based on a number of existing cases suggest that in the oceans, the isostatic depression under the volcanic load extends over a distance generally less than 150 km from the load barycentre (200 km for the huge volcanic mass of Hawaii), and the peripheral bulge extends for between 150–200 and 300–350 km (McNutt and Menard 1978; Pirazzoli 1995). In Hawaii, which is still an active volcano, the rate of subsidence has been estimated at 4.8 mm/yr using tide gauge data (Moore 1971) and at 2.4 mm/yr from submerged terraces

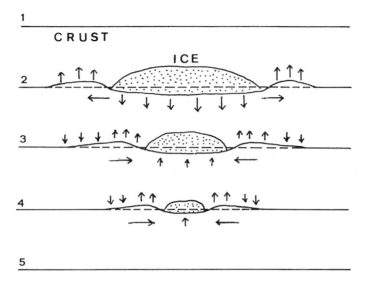

Figure 4   Vertical movements associated with deglaciation (from Daly 1934)

(Campbell 1986), while bathymetric maps show that an arch, 600 ± 200 m high, has developed around the island (Walcott 1970b). In Tahiti, where volcanic eruptions stopped in the mid-Quaternary, subsidence velocity has been estimated at 0.4 mm/yr for the last 100 ka (Montaggioni 1988) and at 0.15 mm/yr for the late Holocene (Pirazzoli and Montaggioni 1985).

In coastal regions, sediment accumulation is noteworthy not only in coral reefs, but especially near major delta areas, where subsidence rates, normally a few millimetres per year (*sediment-isostasy*), can be increased one or two orders of magnitude by extraction of underground fluids (water, oil, gases).

During deglaciation, meltwater from ice sheets produces a considerable load (of the order of 100 t m^{-2} for a sea-level rise of 100 m) on the ocean floor, so that the sea bottom subsides (*hydro-isostasy*). In the upper part of gently sloping continental shelves, or in shallow seas where the postglacial water depth is less than the global change in sea level, the meltwater load will vary according to the local topography (Bloom 1977), generally increasing gradually towards the open sea. In this case the hydro-isostatic constraints will produce a lithospheric flexure with a typical seaward tilting. Potentially good places to measure the glacio-eustatic component of sea-level rise can therefore be found on the outer continental shelf, or on small islands that rise steeply from the deep ocean floor like a dipstick thrust into it. As discussed by Bloom (1971, p. 371), "these places will be warped downward by the water load around them, but because the entire deep ocean floor is depressed, the volume of the ocean basin increases and sea

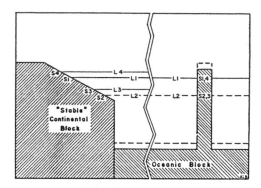

Figure 5   Hypothetical shorelines on a continental coast and an oceanic island (after Bloom 1967). (1) Initial interglacial sea level L1 produces shoreline S1 on the island. (2) Rapidly lowered sea level to L2 produces shoreline S2. (3) Isostatic rebound of the ocean floor carries the island upwards with it, and although the ocean surface rises by isostasy to L3, the island rises too, and there is no change of level between shorelines S2 and S3 on the island. (4) Rapid postglacial rise of sea level to L4 brings the sea to shoreline S4 on the island, which coincides with S1. (5) Slow isostatic sinking of the sea floor lowers sea level to initial position L1, but lowers the island the same amount, so that shorelines S4 and S1 remain superimposed

level with reference to an island, or to a hypothetical buoy moored in deep water, should not change because of the isostatic deformation" (Fig. 5). In the case of an oceanic island, Nakada (1986) has shown that this statement is valid especially when the radius of the island is less than 10 km, and that for a radius larger than 10 km the rheological structure of the upper mantle may modify isostatic adjustments.

Some isostatic movements are immediate, in response to loading and unloading. However, because of the viscosity of the Earth's subcrustal material, movement may continue for several thousand years after loading and unloading have stopped. This is clearly visible in areas of former continental ice sheets, where uplift movements reaching the order of 10 mm/yr are still active several millennia after the ice sheets have completely disappeared.

Several geophysical models have been developed during the last two decades to predict relative sea-level changes following a deglaciation, taking into account the glacio-isostatic and hydro-isostatic components. These models are based on the mathematical analysis of the deformation of a viscoelastic Earth produced by surface loads. The Earth's interior is assumed to be radially stratified, with an elastic structure and a rheology in which the initial response to an applied shear stress is elastic but the final response is viscous. Having assumed a melting history for all the continental ice loads that existed at the time of the last glacial maximum, the meltwater distribution within and among the oceans is constrained in a way

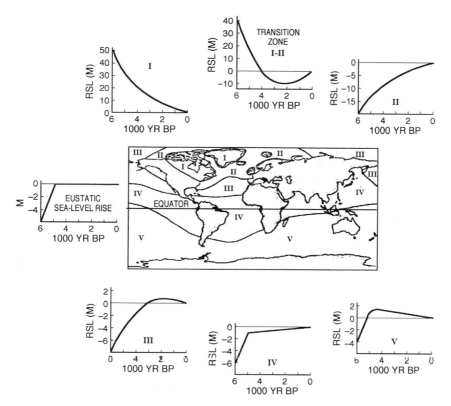

Figure 6   Sea-level zones and typical relative sea-level curves deduced for each zone
by Clark et al. (1978) under the assumption that no eustatic change has occurred
since 5 ka BP (adapted from Clark and Lingle, 1979)

such that the instantaneous geoid remained an equipotential surface at all
times. These models have demonstrated that the rate, direction and magnitude of crustal movement must have varied from place to place and
therefore that no region can be considered as vertically stable (Fig. 6). They
have also been useful in providing a first-order approximation of the
deglacial and postglacial sea-level history in areas where no data are
available.

1.3.2   Other causes of local vertical deformation

Local sea-level changes can have several other possible causes. In
tectonically active areas, neotectonic deformation is common and may be
due to stresses related to isostatic processes, but also to compression,
gliding, elastic rebound, faulting, folding, or tilting of crustal blocks.

Trends of vertical displacement of tectonic origin often appear to be continuous and gradual over the long term, but frequently consist of spasmodic movements, often occurring suddenly at the time of earthquakes of great magnitude. Rapid aseismic displacements have also been reported. Active folding is frequent, and it is detected generally on the basis of geomorphological and geodetic observations. Methods of neotectonic inquiry depend on the time scale investigated (Vita-Finzi 1986; Stewart and Hancock 1993): seismological data and geodetic data for short periods (from less than a day to 100–1000 years), historical, archaeological and geomorphological data for periods from one day to over 1 Ma, and geological data from 10 ka to over 10 Ma. All these methods can be of use for the last 20 ka.

Displacements occurring at the time of an earthquake are called *co-seismic*. Gradual displacements, often reversing the coseismic ones, may occur during a few years or decades preceding (*preseismic*) or following (*postseismic*) the coseismic event. The average interval between two coseismic events (*interseismic*) can vary, depending on the seismo-tectonic area considered, from a few centuries to over 10 ka (see Section 5.3). The repetition of coseismic uplift or subsidence during a period of relatively stable sea level will produce sequences of stepped shorelines, which have been reported from many seismic coastal areas for the Holocene period. Similarly stepped shorelines could, however, be produced by sea-level fluctuations (Fig. 7), and misinterpretations should be avoided.

Compaction of sediments depends on changes in their water content, following emergence and drainage, loading pressure, or pumping from underground layers (see Section 5.4). In delta areas, like the Mississippi, for example, contemporary silty muds often contain water to 50% of their weight. In peat layers, compaction may reach as much as 90% (see Chapter 2). Recent human-induced subsidence can be important in river-mouth or lagoonal areas and land sinking has been reported from many coastal regions: 4.6 m in Tokyo (Japan), 2.7 m in the Po Delta (Italy) (see Section 5.4), 2.7 m in Shanghai (China) and Houston (USA).

1.4 VARIATIONS OF WATER DENSITY AND DYNAMIC CHANGES AFFECTING THE WATER MASSES

1.4.1 Steric changes

Even if the volume of the oceanic basin and the quantity of sea water remain constant, sea level may change owing to variations in sea-water density, which depends on salinity, temperature and pressure: it decreases when temperature increases, and increases with salinity and pressure, the latter increasing with water depth. Denser water occupies a smaller volume,

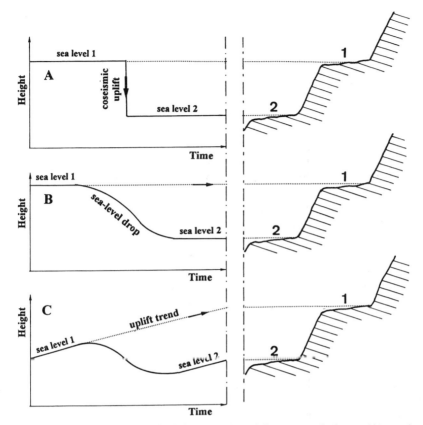

Figure 7   Stepped morphological features 1 and 2 may result from: (A) sea-level stability and coseismic uplift, (B) a gradual sea-level fall, and (C) a sea-level fluctuation superimposed on a gradual uplifting trend

so that sea level drops; a decrease in water density causes a sea-level rise. To give an order of magnitude, an increase of 1°C over 4000 m would produce a rise of 60 cm. A variation of 4‰ in salinity has the same effect as a variation of 1°C. Such salinity variations are exceptional in mid-ocean, and most steric changes are caused by thermal variation. However, in coastal areas and estuaries, freshening may have a substantial effect on the steric level of the sea.

During the last glaciation, when enormous fresh-water masses were stored in ice sheets, the mean salinity of the oceans was about 3.5% above present, i.e about 36‰, thus comparable with the southern Red Sea today (Fairbridge 1961). The average water temperature was also slightly colder than today, so that sea water was denser.

The main gravitational forces (the attractive force of the Earth and the centrifugal force of the Earth's rotation) tend to level the differences in atmospheric pressure and water density created between the equator and the poles by the uneven distribution of heat over the Earth's surface. This causes oceanic currents and dynamic changes in sea level. MSL differences due to density and dynamic changes have been called *steric changes* (Pattullo et al. 1955). These changes are measured as differences in dynamic level above isobaric surfaces, i.e. above imaginary surfaces along which pressure remains constant. A dynamic metre (dyn. m) has the dimensions of work and represents the work which must be done to move a unit mass the distance of one metre in the vertical direction. It is determined by the potential value gh (gravity × vertical distance). Choosing the dynamic metre as a unit for the dynamic depth, we obtain $D = gh/10$. One dynamic metre therefore corresponds roughly to 1.02 geometrical metres. At the present time, steric variations produce differences in sea level as large as 2.6 m, between a high (4.00 dyn. m) south of Japan and a low (1.40 dyn. m) in the South Atlantic, near Antarctica (Fig. 8). It can be estimated from Fig. 8 that the difference in steric heights which may be produced between two oceanic sites by a climatic change, like those which occurred in the Quaternary, is not more than a few metres. Such changes may, however, occur over relatively short distances (Figs 9, 10) and in short periods (Fig. 111).

1.4.2 Tidal changes

The tide also depends on gravitational forces. There is a close relationship between the tidal constituents in a point of a basin and the shape of the same basin. Any modification of the basin morphology will affect the values of the tidal constituents. Sedimentary accumulation or erosion, land subsidence, tectonic uplift or tilting, sea-level changes, and, more recently, human interventions, can all modify the shape of a basin, and therefore its tidal range.

Postglacial sea-level changes have certainly been an important cause of tidal variations, especially in macrotidal regions. In the Bay of Fundy, which experiences the largest tidal range in the world (up to 16 m) (Stea 1987), high-tide level has risen fairly steadily at about 3 mm/yr for the last 4 ka, twice the rate on the nearby Atlantic and Gulf of St Lawrence coasts. The Bay of Fundy has submerged up to 6 m more than the rest of the Maritime Provinces in postglacial times, when measured with reference to a high-tide datum, clearly showing a gradual increase in the tidal range with sea-level rise (Grant 1970). In The Wash area (England), by using an approach that integrates numerical tidal models with stratigraphic data recording former tidal heights and palaeogeographic simulations, Hinton (1992) has estimated that changes greater than 1.5 m in the mean high-water spring tide (MHWST) level occurred in the inner part of the bay during the last 4 ka; in general there has been an overall increase in MHWST altitudes in The Wash with sea-level rise towards the present day (Hinton 1995).

19

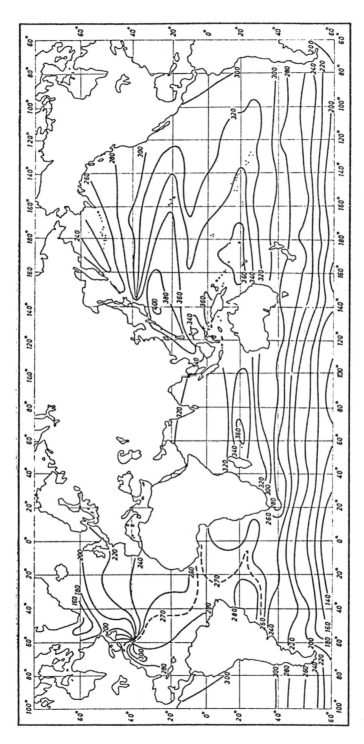

Figure 8   The distribution of mean sea-level altitude (dyn. cm) in the world ocean (after Lisistzin, 1965, 1974)

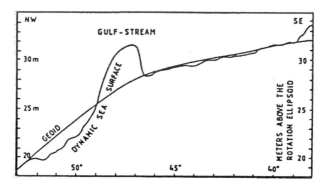

Figure 9   Actual satellite profile of the sea-surface topography in the western North Atlantic indicating how the strong current forces of the Gulf Stream force the dynamic sea surface to deviate from the geoid level, according to Mörner (1994)

GEOS-3 COLLINEAR ALTIMETER DATA

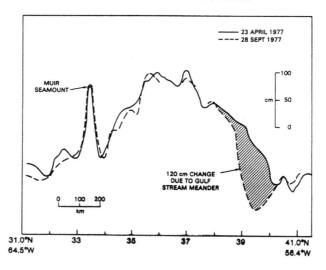

Figure 10   Two Geos-3 altimeter profiles crossing the Muir seamount north of Bermuda show an identical geoid undulation at the position of the seamount, with, however, a change in height (shaded area) due to a meander of the Gulf Stream (after Douglas et al. 1983)

Since the 16th century it has been known that exchanges of tidal waters between coastal lagoons and the sea are controlled, up to a certain limit, by the ratio, S/I, between the surface area of the lagoon (S) and the cross section of the inlets (I). When I increases, sea water penetrates more easily into the lagoon. On the other hand, when S decreases, sea water penetrating through the passes will spread further into the lagoon more quickly

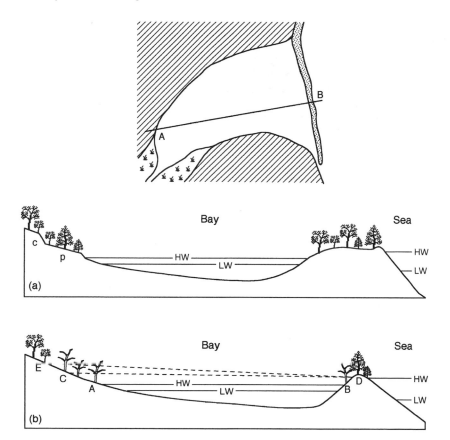

Figure 11   Tidal changes within an open lagoon resulting from formation of an enclosing spit (a) or enlargement of the lagoon entrance (b)

and the tidal range will be larger, sometimes reaching levels close to or even higher than those prevailing in the open sea. A classic example of apparent sea-level fluctuations due to tidal changes was published by Johnson (1912), who considered the common case of a funnel-shaped estuary transformed into a lagoon by the development of a littoral bar at its entrance (Fig. 11). In the inner part of the lagoon, the tidal range will be smaller than in the open sea, with apparent emergence of the cliff (c) and of the erosion platform (p) cut by the sea before the development of the bar, when the estuary was open to wave action (Fig. 11a). If a storm breaks the littoral bar, or if the section of the lagoon pass is increased artificially (e.g. for navigation purposes), the tidal range inside the estuary will increase, the high-water level passing from A to C, killing pre-existing vegetation and creating an impression of submergence (Fig. 11b). Because the tidal range in a funnel-shaped basin usually becomes greater than on the open coast, marks of apparent sea-level rise will reach higher levels (E) in the inner part of the estuary than on the open coast.

Increases in the tidal range due to the deepening of navigation channels have been reported from several coastal areas, e.g. from the New York Harbour (Marmer 1943), the lagoon of Venice (Pirazzoli 1975, 1987b) and in several estuaries.

Chapter Two

Evidence of former sea levels

Present sea level is producing many features indicative of its position, some of which may be preserved for a long time after this position changes. This chapter is devoted to a description of sea-level indicators which may be used, when preserved, to recognize former sea levels.

2.1 ROCKY SHORES

2.1.1 Marine zonation

The most common sea-level indicators are marine organisms and communities which are naturally arranged in series of horizontal bands (marine zonation). In the field, evidence that a change related to particular tidal horizons has taken place often appears from a comparison between the present-day bands (with living organisms) and similar fossil organisms which now stand at a different level. There are also features produced by bioerosion at present and past sea levels. On rocky shores (Stephenson and Stephenson 1949, 1972) a number of biological zones have been identified in which bioerosive and/or bioconstructive forces are active.

(1) The *littoral fringe* (Stephenson and Stephenson 1949) or *supralittoral zone* (Pérès and Picard 1964) is never submerged, but is kept wet by spray. Its upper boundary is the position beyond which sea spray ceases to be an effective force in determining community distribution. The upper part of the supralittoral zone is characterized by lichen vegetation and is maritime rather than marine in character, since the vegetation consists mostly of salt-tolerant species with marked terrestrial affinities. The lower supralittoral zone, which is more distinctly marine in character, is typically dark-grey in colour, the blackening being due to the presence of lichens such as *Verrucaria*, or *Lichina*, and/or of associated Cyanobacteria (blue-green algae). The importance of the latter is usually diminished in subpolar regions, but becomes predominant in warm temperate, subtropical and tropical regions. On carbonate substrates, plant roots create many microtunnels which are

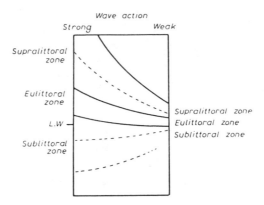

Figure 12   Explanatory diagram of zonal terminology adopted in the text. Three primary zones are recognized and their boundaries indicated by continuous lines. Dashed lines denote possible subzonal boundaries in supralittoral and sublittoral zones. Note zonal enlargement and displacement due to effects of increasing wave action. LW indicates the approximate position of low water on both tidal and non-tidal shores (after Russell 1991)

easily preserved and fossilized (Le Campion-Alsumard 1979–1980). Among these boring microorganisms, the presence of Cyanophyceae (endolithic algae) is easy to recognize in the darker part of limestone cliffs above the high-tide level, since they are the only plants able to tolerate the extreme conditions of such an environment. Other algae which may occur in the lower supralittoral zone are the green *Endoderma* or brown Crysophyceae, which may also be overgrown by more conspicuous algae such as *Bangia, Porphyra* (red algae), *Ectocarpus* (brown algae) and various green algae. The most widespread animals in the lower supralittoral zone are the gastropods *Littorina* and *Melanerita*, which graze algae, the crustaceans *Ligia* and *Megaligia*, which are detritus eaters, and other more occasional opportunistic visitors which do not normally reside there, such as hermit crabs, insects, spiders, birds and small rodents. Bioconstruction is unusual in this zone, but it occurs where there are occasional incrustations by *Chthamalus* shells in shadowed rock crevices exposed to wavesplash.

(2) The *midlittoral (eulittoral) zone* is frequently submerged by waves and tides. Its upper limit may be recognized from certain boundaries, for example the "litus line" (the boundary between *Pelvetia canaliculata* and *Fucus spiralis*) in Norway (Russell 1991). On tidal shores, the lower boundary of this zone lies at, or a little above, mean low spring tide. However, the limits of the midlittoral zone, like those of the supralittoral zone, are raised and widened with increasing wave action (Fig. 12). The upper part of the midlittoral zone is never rich in species. On Arctic and Antarctic shores it is barren for most of the year, owing to the grinding action of ice, and may become vegetated by benthic diatoms and ephemeral

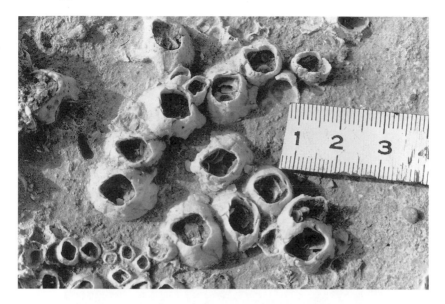

Figure 13   Fossil barnacles left about 2.5 m above their live counterparts on the north coast of Qeshm Island, Iran. They have been dated by M. Fontugne to 6210 ±50 years BP (Gif-9848). Scale is 4 cm. Local tidal range is about 5 m (Photo E158, May 1994)

algae only during summer periods. *Fucus* and *Pelvetia* algae are often present in North Atlantic and North Pacific coastal areas. Further south there is a greater proportion of annual species and the importance of Cyanobacteria increases when the tropics are reached. Lithothamnia species have been reported from exposed shores of the Indian Ocean and the Pacific, as well as such species as *Cladophora, Ectocarpus* and *Ulva*.

The lower part of the midlittoral zone usually has a dense cover of fucoid and turf algae, which attracts a large number of herbivores (littorinids, limpets).

Barnacles (Fig. 13), mussels and oysters are often present in the midlittoral zone. Barnacles (*Balanus, Elminius, Tetraclita* and especially *Chthamalus*) are found in the upper part of this zone, whereas mussels (*Mytilus*) and oysters tend to occupy lower levels. Organisms which may be fossilized *in situ* after death include barnacles, mussels, oysters and, in the almost tideless western Mediterranean, *Lithophyllum lichenoides* (also known as *Lithophyllum tortuosum*), which forms a rim just above sea level (Laborel 1987). Erosive agents are important: Cyanobacteria in the upper part of this zone, limpets (*Patella* spp.) and Chitons in the lower part, together with the pelecypod *Lithophaga lithophaga* (Laborel and Laborel-Deguen 1994) contribute actively to shape the intertidal erosion bench and the tidal notch, especially on limestone coasts.

Figure 14   The calcareous alga *Neogoniolithon notarisii* (Dufour) has a narrow
zonation in the uppermost part of the sublittoral zone and is therefore an excellent
sea-level indicator (south coast of Crete, Greece). Local tidal range is less than
0.3 m, and water level is at the trough of a wave (Photo 4227, Sept. 1977)

(3) The *sublittoral (infralittoral) zone* extends from its boundary with the
midlittoral zone down to a depth which may reach 25–50 m, depending on
water transparency. This depth, at which growth of macroalgae ceases
(Russell 1991), is very variable, and the sublittoral zone may be absent
altogether in very turbid coastal waters. Otherwise, this zone is densely
populated by brown algae, coralline encrusting algae (*Porolithon, Neogoniolithon, Lithophyllum*) (Fig. 14), fixed vermetid gastropods (e.g. *Dendropoma
petraeum, Vermetus triqueter, Serpulorbis arenarius*), Cirripeds like *Balanus*
spp. or coral reefs in warm waters. They may be accompanied by turf-
forming algae and fucoid vegetation (*Fucus, Cystoseira, Sargassum*). These
are grazed by herbivorous fish and sea urchins; the latter, together with
Clionid boring sponges and other borers, may also attack a rocky substrate.
Organisms which may be fossilized *in situ* in the sublittoral zone include all
bioconstructions (encrusting algae, vermetids, reef-building corals and
associates, barnacles, oysters). The best altimetric indications for the reconstruction of former sea levels are obtained from comparison of the fossil
bioconstructions with their present-day counterparts; especially useful is the
identification in the fossil record of those bioconstructions which have the

Figure 15   The outer rim of these pools is formed by *Dendropoma* and other vermetids which are regularly wetted by waves on the front of the fringing reef of Tubuai, Austral Islands, French Polynesia. Local tidal range is 0.6 m, and water level is near MSL (Photo 7855, Oct. 1983)

narrowest vertical zonation, e.g. the upper level of coral microatolls, of pools made by *Dendropoma* (Fig. 15) or *Neogoniolithon*, the algal crest of a former coral reef (see below).

2.1.2   Erosional indicators

Geomorphological features may be erosional or depositional. Erosional indications can be preserved only in hard rock, and occur in a vertical range which depends on site exposure. Erosional features which may be useful as sea-level indicators include notches, benches and platforms, pools, potholes, sea caves, honeycombs and certain bioerosion features.

Several kinds of erosional notches have been classified by Pirazzoli (1986a). *Structural notches*, which depend on differential erosion in weaker rock layers rather than on sea-level position, should be recognized as such and not used as sea-level indicators. *Abrasion notches* may develop at any level reached by wave action, provided that granular deposits are available to be projected against the cliff face. In macrotidal areas abrasion notches are most frequent near high-tide level, or at higher elevations on exposed coasts; in microtidal areas, however, active abrasion notches have also been reported underwater, as near Tottori (Japan), where a notch sloping between +0.5 m and −2.0 m has been carved on granite cliffs bordered by a beach (Toyoshima 1965). The term "wave-cut", often used in the literature,

Figure 16   This tidal notch at Itanma, Okinawa (Japan), was raised more than 2 m at the time of an earthquake dated 2350–2400 years BP. Local tidal range is 1.6 m, and water level is 0.55 m below MSL (from Pirazzoli and Kawana 1986) (Photo 6634, March 1981)

Figure 17   Small tidal notch cut into the limestone cliff at about +2.9 m near Cape Teodoco, Rhodes Island, Greece, indicating the highest level reached here by the Holocene sea. Local tidal range is less than 0.3 m (Photo 4561, May 1978)

is correct when applied to structural or abrasional notches, but should not be used for *tidal notches*, which are one of the more precise sea-level indicators, since mechanical wave action is not instrumental in cutting these notches (Figs. 16, 17).

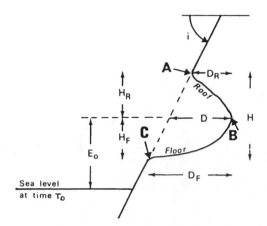

Figure 18   Terminology of a tidal-notch profile: C, base; B, retreat point; A, top; H, height; D, depth; E, elevation; i, cliff slope; H_R (H_F), roof (floor) height; D_R (D_F), roof (floor) depth; T_o, time of measurement

Tidal notches are typical midlittoral erosional features, especially in limestone coasts. With a moderate tidal range, the most common profiles are recumbent V-shaped or U-shaped forms. In a sheltered site the depth of a tidal notch increases gradually above the notch floor, situated near the lowest tide level, to a segment with maximum convexity which usually occupies the neap tide range. The *retreat point* or *vertex* is located near MSL. From that point the depth decreases gradually upwards and becomes negligible at the top of the notch, near the highest tide level (Fig. 18). In a moderately exposed site, where wave action regularly splashes sea water above the high-tide level, the top will tend to be shifted upwards and the height of a tidal notch will increase with exposure; in very exposed sites, however, tidal notches may be lacking altogether.

According to this vertical zonation, which has been described from many places, the development of tidal notches appears to be most active where the intersections of (limestone) rock, air and sea are frequent and regular, and is seen to decrease abruptly in the supralittoral and sublittoral zones. Though early studies ascribed this zonal erosion to solution phenomena (for references, see Pirazzoli, 1986a), it is now recognized that tidal-notch development is mainly related to bioerosion (Trudgill 1976; Torunski 1979) and zones of rapid tidal-notch development correlate well with zones of major eroding organisms, the maximum density of grazing organisms being near MSL. Rates of undercutting of 0.2 to 5.0 mm/yr have been reported in the literature, the most frequent values being of the order of 1.0–1.5 mm/yr in the tropics and possibly much less on hard Mediterranean limestones. The period necessary for the development of a tidal notch in a cliff profile increases with tidal range, and when the cliff face is not vertical (Pirazzoli 1986a).

Figure 19 Double erosion bench near Caraci (Haraki), Rhodes Island, Greece. The upper bench corresponds to a former higher sea level; the lower bench is related to the present sea level. Local tidal range is less than 0.3 m (Photo 4588, May 1978)

Not to be confused with tidal notches are *surf notches*, which may occur on limestone coasts exposed to persistent winds and strong surf and spray action. Surf notches are only developed above a surf algal bench (Focke 1978) (see below).

Erosional *benches* and *platforms* are often used as former-sea-level indicators (Fig. 19). The formation of coastal benches and platforms is usually ascribed, outside coral reef areas, to the removal by waves of the weathered parts of cliffs. The lowest level of possible weathering probably corresponds to that of constant soakage by sea water, which is placed by most authors in the intertidal zone. Wave abrasion is effective on most rock formations, especially in structurally weak layers, and may occur at any elevation between the highest levels reached by storm waves and the maximum depth at which sediments of the sea floor are moved by waves. Most often, when a cliff is retreating as the result of wave action, it leaves an erosion platform sloping gently seawards, between near high-tide level at the cliff base, and an outer level which is usually lower than the local low tides. If the rock is too hard, however, no platform will develop, and if it is too weak, frequent slumping will prevent the evolution of a cliff.

An abrasion notch (Fig. 20) is often developed near high-tide level at the boundary between the cliff and the shore platform. The average elevation of shore platforms usually increases with rock hardness; certain platforms may also develop well above high-tide level when water supplied by waves can stagnate behind an outer rampart.

Figure 20   Abrasion platform cut into a volcanic breccia by a higher Holocene sea level near Ching-pu, on the east coast of Taiwan. An abrasion notch is visible on the left. Local tidal range is 1.4 m (Photo B777, Jan. 1990)

On limestone coasts exposed to persistent winds and to strong surf and spray action, organic accretions often develop on the notch floor, or near the outer edge of the bench, protecting the substrate rock and thus locally inhibiting further erosion. While erosion proceeds above the accretion level, a *surf bench* will begin to form, and may become several metres wide (Fig. 21). Surf benches have been reported as high as 2 m above high-tide level. However, their level declines with decreasing wave action (Focke 1978). On the eastern Mediterranean coasts, surf benches (there called *trottoirs*) are quite common, rimmed by encrusting vermetids and algal accretions. As surf action there is relatively moderate, trottoirs are usually found no more than 0.2 to 0.4 m above MSL (Figs. 22, 23).

In high latitudes the weathering effects of freeze–thaw alternations, in addition to wave action, are thought to contribute to the formation of *strandflats*, which are low shore platforms where disintegrated rock material, resulting from freezing and thawing, is readily washed away by wave action. In limestone areas, the effects of strong bioerosion in shallow water may also contribute to platform formation, and intertidal lowering rates of 0.4–2.5 mm/yr for limestone platforms were reported from Kaikoura Peninsula (New Zealand) by Kirk (1977).

In short, the reliability of erosional coastal platforms or benches as sea-level indicators depends on the identification and understanding of the processes which were active in the development of these coastal features.

Figure 21   Contemporary surf bench cut into Quaternary reef limestone at 0.5 m
above the high tide level near Cape Ara, Okinawa, Japan. Local tidal range is 1.7 m,
and water level is 0.3 m below MSL (Photo 6580, Feb. 1981)

Figure 22   Contemporary surf bench near Dor, Israel. Local tidal range is less than
0.3 m (Photo A357, Sep. 1986)

Figure 23   Emerged surf bench corresponding to a relative sea level 0.6 m above the present level, near Ras al-Bassit, Syria. The uplift has been dated from about 1400 conventional years BP (Dalongeville et al. 1993). Local tidal range is less than 0.3 m (Photo D331, Aug. 1991)

Honeycombs are small cavernous features assuming a cell-like structure (Fig. 24); they have been reported as occurring in a variety of medium- to coarse-grained rocks, such as sandstones, granites, lavas, conglomerates, gneiss, schists and limestones. They generally form in the spray zone above high-tide level and their development may be ascribed to temperature variations, chemical weathering (including recurrent wetting and drying, and salt crystallization) and wind corrasion. When fossilized, they may indicate the proximity of sea level below them, with an accuracy which can be very poor on exposed sites. *Tafoni* are cavernous weathering features of greater size (from a decimetre to several metres), developed on the rock surface. They do not occur only in coastal areas, and therefore are of no use as sea-level indicators.

Coastal *pools* are flat-bottomed depressions frequently found on lime-stone benches. In the midlittoral and supralittoral zones, they may result from solution and/or bioerosion processes; these may be used as good indicators of past sea levels on sheltered coasts, whereas on exposed shores they only indicate supralittoral areas in the reach of waves. Coastal pools should not be confused with similar features partly bioconstructed by vermetids or by calcareous algae at the uppermost limit of the sublittoral zone (which are very accurate sea-level indicators), or with underwater basins derived from coalescence of sea-urchin niches.

Marine *potholes* are rounded depressions, generally with greater depth than width, worn into the solid coastal rock by sand, gravel, pebbles and boulders being spun round by the force of the waves. On exposed coasts, potholes may be formed at various levels, ranging from some metres above high tide to well below low tide, and so are not good indicators of past sea level.

Figure 24   Honeycomb structures developed on a sandstone formation just above the high-tide level at Kannoura, Muroto Peninsula, Shikoku, Japan. Local tidal range is 1.5 m (Photo 359, Mar. 1974)

Sea caves are hollows excavated by marine erosion into a cliff in the range of wave action (Fig. 25). They generally occur where the weaker parts of a rock formation (soft layers, joints, faults, breccias, shale beds, unconformable strata, irregular sedimentation, internal structures of lava flows) have been excavated by the sea. In limestone formations, sea caves are often of karstic origin, enlarged and reworked by the sea. A sea cave open to the sea on both sides of a promontory is a *sea arch*. Sea caves and arches have generally little value as sea-level indicators; in limestone formations, however, the floor of a sea cave or an arch, if regular and flat, may be related to a former low-tide position.

Several bioerosion features may provide indications of former sea-level positions. Marks left by supralittoral plant borers will indicate a formerly lower sea level, usually with poor accuracy. Borer shells, on the other hand, which can usually live at any depth in the sublittoral zone, will indicate a higher sea level, and the same can be said for erosional features left by boring sponges or sea urchins (Fig. 26). However, when the elevated upper limit of burrows (e.g. of *Lithophaga* in the Mediterranean) is clearly defined and forms a horizontal line, this corresponds to a former MSL (Stiros et al. 1992; Laborel and Laborel-Deguen 1994) (Fig. 27), which can even be dated by radiocarbon assay when the articulated shells of the burrowing animals have been preserved inside the burrows.

Figure 25 Sea caves cut into limestone cliffs at Tripitis, south coast of Crete, Greece. Their floors correspond to a higher sea level and they were uplifted coseismically in AD 365 (Photo 4296, Sep. 1977)

Figure 26 Emerged holes excavated by sea urchins into doleritic rocks during a slightly higher late Holocene sea-level stand. Cape Almadies, Sénégal (Photo 5283, Nov. 1979)

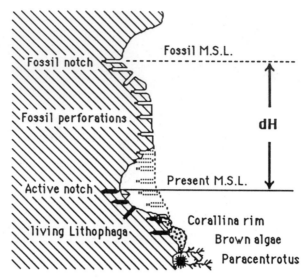

Figure 27   Measurement of a former elevated sea level using the past and present upper limits of *Lithophaga* burrows: dH, difference in elevation (adapted from Laborel and Laborel-Deguen 1994)

2.1.3   Depositional indicators

Distribution of *coral reefs* depends mainly on sea temperature (usually between 17–18°C and 33–34°C), salinity (normally in a range from 30 to 38‰) and water turbidity (Guilcher 1988). Owing to light requirements, coral reef growth is limited to the sublittoral zone.

A coral reef consists of scleractinian corals, which, with crustose coralline algae (Melobesiodeae), are the main framework builders, associated with many other living organisms (such as foraminifera, crustaceans, gastropods and lamellibranchs) and with sediments (generally carbonates) trapped in the framework and finally cementing it. The uppermost limit of coral growth is determined by emersion and is normally close to mean low-water spring tide level (Hopley 1986b). Some species among the reef builders have a narrow vertical zonation, or are typical of shallow-water environments; they can therefore be employed as useful sea-level indicators. When only cored samples are available, on the other hand, some reservations and difficulties may arise in environmental reconstruction, owing to doubts on the growth position in the cores, because reef debris may be removed by wave action and deposited as rubble in other reef zones (Fig. 28). Hopley (1986b) estimates that less than 5% of the reef structure in northern Queensland is formed from *in situ* coral. On the other hand, certain reefal morphological features (reef flat, algal crest, microatolls) can be related to sea level with quite narrow uncertainty ranges.

Figure 28   *In situ* outcrop of an emerged coral reef dating from 8–6 ka BP near Qwambu, Huon Peninsula, Papua New Guinea. Only a few corals are clearly in growth position. John Chappell gives scale (Photo A894, Aug. 1988)

A *reef flat* represents the top of the lithified part of a coral reef which has reached the sea surface (at low tide). It consists of an almost horizontal pavement cemented by calcareous algae, which may reach higher levels after having been capped by lithified detrital elements derived from the coral community (coral reef conglomerate) and by sand cays. In elevated reefs, a discontinuity, consisting of an almost horizontal fissure, may mark the boundary between the top of the bioconstructed reef and the conglomerate capping it (Fig. 29). The width of a reef flat may vary from less than 100 m to several kilometres. Reef flats are often totally exposed at low spring tides, although examples of reef flats which may remain partly submerged all the time are also known.

Individual corals that have not developed into a coral reef can also be used as sea-level indicators if their vertical zonation is known.

Figure 29   A slightly emerged continuous fissure (arrow) marks the boundary between an ancient reef flat with corals emerged in growth position and a conglomerate flagstone developed above them on Temoe Atoll, Gambier Islands, French Polynesia (Photo 7459, Oct. 1982) (from Pirazzoli 1987a)

Microatolls are individual subcircular coral colonies (generally but not exclusively of *Porites*) with a flat upper surface devoid of living polyps, not exceeding a few metres in width (Fig. 30). A slightly raised outer rim may exist above a dished centre in the flat surface. Microatolls have been described in detail by Scoffin and Stoddart (1978), and provide very accurate sea-level indications. The shape of a microatoll can only be due to a limitation in the upward growth of the corals. This limitation is probably related to emergence, which causes desiccation of the polyp mucus after about 3 hours of exposure in high day-time temperatures. According to Hopley (1986b), the uppermost level of living coral growth on the Great Barrier Reef approximates mean low-water spring tide whatever the tidal range, with a variation of up to 25 cm owing to year-to-year change in tidal behaviour, varying exposure to wave action, or varying tolerance to exposure of different species. In some areas (e.g. in Sumba Island, Indonesia) lowest spring tides may occur only at night, when desiccation rates are slower, thus enabling corals to endure longer emergence and survive at slightly higher elevations.

In some cases, water may be temporarily prevented from draining to the ocean during the falling tides, held back in part of the reef flat by some kind of barrier (e.g. by a rampart of deposits left by storms). Behind such a dam, the tide will fall to the same level at every tide, a level independent of the openwater low tide. In this case the upper level of coral growth (and of microatoll development) will be higher than in the ocean as long as the outer rampart protects it against wave action. Slightly emerged microatolls

Figure 30 Fossil *Porites* microatoll, corresponding to a slightly higher late Holocene sea level, denuded when the sediments capping it were carried away by the 1983 cyclones. A double metre rule (folded) gives scale. Tupai atoll, French Polynesia (Photo 7934, Oct. 1983) (from Pirazzoli et al. 1985a)

should therefore be interpreted with caution, taking account of the possible variations in coastal topography which may have caused such moating.

At the outer limit of a coral reef, an algal coating (*coralline algae*) with *Porolithon* may form a ridge or a broad irregular dam, 5 to 15 m wide, rising a few decimetres above the reef flat (Fig. 31). This algal crest, which is especially well-developed on the windward side of a reef, is an excellent datable indicator of former higher sea level when recognized in fossil counterparts. On the outer edge of the reef, down to a few metres, *Porolithon* may be found forming more or less spaced downslope grooves, separated by spurs (Fig. 32), which may remain well-preserved and easily recognizable in elevated reefs (Fig. 33) and thus help to estimate the amount of emergence.

Coralline algae have a wider distribution than coral reefs, occurring along coasts not dominated by large stretches of shifting sand and where salinity is not abnormal. In subtropical to temperate regions, the reef frameworks developed by these algae are much smaller than in the tropics, but still potentially preservable in the fossil record. In boreal and arctic waters, lithification of the framework rarely occurs, and destructive elements exceed constructive ones. Nevertheless, in cold waters, coralline algae often provide a subcontinuous crust over a rocky substrate, and in some cases this crust can be preserved in place (Adey 1986). Single fossil plants provide little information regarding sea-level position at the time of growth, but the depth

Figure 31   Algal crest interrupted by transverse surge channels west of Motu Auira, Maupiti Island, French Polynesia. Local tidal range is less than 0.3 m (Photo 7612, Oct. 1982)

Figure 32   Contemporary spurs and grooves seen underwater from the air at the outer edge of the Tahiti reef flat, French Polynesia (Photo 7225, Sep. 1982)

at which fossil material grew can be estimated on the basis of community structure where coralline organisms are abundant in a Quaternary reef framework. It is possible to determine former sea level within ±2 m near mean low water, within ±4 m from 5 to 20 m, and within ±15 m at depth greater than 30 m (Adey 1986). On the Mediterranean coast, precise sea-level indications can be provided by two calcareous algae: *Lithophyllum lichenoides* and *Neogoniolithon notarisii*. *L. lichenoides* is found mainly in the

Figure 33   Spurs and grooves extending above sea level in Toku Island, Ryukyus, Japan, give evidence of Holocene emergence (Photo 208, Mar. 1974)

western Mediterranean, where it may form a calcareous rim or a wide overhanging corniche with a flat upper surface in the lower part of the midlittoral zone on cliffs, in coves and in crannies moderately exposed to wave action (Laborel 1987; Morhange 1994) (Figs. 34, 35). Submerged rim remains of *L. lichenoides* have been recently identified and dated at several sites, providing evidence of low sea-level histories (Morhange 1994; Laborel et al. 1994). *N. notarisii* is more common in the warmer waters of the eastern Mediterranean, where it is found in the upper fringe of the sublittoral zone, and forms, alone or in symbiosis with the vermetid *Dendropoma petraeum*, organic accretions on the outer edge of surf benches or overhanging small corniches on cliffs.

Submerged speleothems from sea-flooded karst systems can also provide useful information, especially when they are capped by a marine biogenic cover preserved from bioerosion in the relatively dark sea-cave environment: the date of the outer part of the speleothem at a certain depth will indicate that the sea level was still at a lower position, whereas the date of the beginning of the marine bioconstruction will approximate the time when the sea rose above that level (Alessio et al. 1992, 1994).

Encrusting shells (barnacles, oysters, vermetids) preserved in growth position are also excellent sea-level indicators. Submerged, thick crusts can be partly preserved, especially if capped by younger bioconstructions or concealed by sedimentation. Thin crusts, on the contrary, are likely to be destroyed by sublittoral erosion as they are submerged. If there is slow

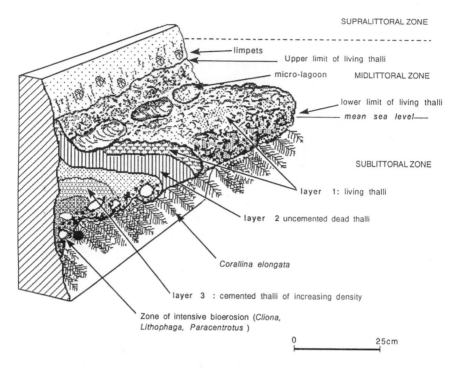

Figure 34   Structure and morphology of a *Lithophyllum lichenoides* rim (from Laborel et al. 1994)

Figure 35   *Lithophyllum lichenoides* rim (arrow) developed on a Cretaceous limestone cliff near Cassis, southern France (Photo E458, Sep. 1995)

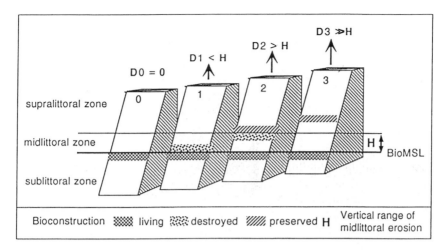

Figure 36   Diagram showing tectonic uplift of a coastal block with subsequent erosion (mainly limited to the midlittoral zone) of bioconstruction belts elevated by vertical upward movements of increasing amplitude (after Laborel and Laborel-Deguen 1994)

emergence (a few millimetres per year), their remains will be carried gradually up through the midlittoral zone, where they will be subject to the attack of limpets and rock-boring Cyanophytes. Small elements, such as isolated vermetid tubes or barnacle shells, will be smothered and will disappear completely in a matter of a few years or decades (Laborel and Laborel-Deguen 1994). They are more likely to be preserved if there is rapid emergence (e.g. after an uplift accompanying a great earthquake) (Fig. 36). Thicker crusts, like algal rims and coral reefs, may be deeply etched by midlittoral erosion, but not completely destroyed; their remains, though partly eroded, are still useful as past sea-level indicators.

Fossil intertidal barnacles have seldom been used to measure and date relative sea-level changes, perhaps because many shells are necessary for a classic radiocarbon dating, and because the collecting and cleaning of these shells is often a time-consuming exercise. With the advent of accelerator mass spectrometry, which requires only a few milligrams of carbon for each dating, the use of barnacles as fossil sea-level indicators may increase. Problems related to finding, sampling, cleaning, dating and interpreting the altimetric significance of fossil barnacles have been discussed by Pirazzoli et al. (1985b). *Chthamalus* barnacles, which live in the splash zone and reach relatively high elevations on exposed coasts and in rock crevices, are considered as poor indicators of past sea level.

More massive than barnacles, oyster shells may form thick incrustations, and certain intertidal species have bioconstructed rims which can be used as precise sea-level indicators (Figs. 37, 38). The calcareous tubeworm

Figure 37   Oyster shells forming a corniche around a limestone block at Antany Mora, Radama Islands, Madagascar. Water level is about 0.2 m below MSL (spring tide range: 3.4 m), and the vertical scale is 2 m. This profile was misinterpreted from a distance by R. Battistini as a double notch indicating Holocene emergence (Guilcher et al. 1958, p. 147 and plate XXIIIC). In reality there is only a single tidal notch, the height of which is consistent with present-day tide range, a bioconstructed corniche projecting in the middle (Photo 8795, Jan. 1986)

Figure 38   Detail of a corniche constructed by oysters, as shown Fig. 37 (Photo 8796)

Figure 39   Cushion-like growths of the calcareous tubeworm *Galaeolaria caespitosa* on pier supports at Rosebud, on the southeastern shore of Port Phillip Bay, Australia, may provide a biological indicator of sea-level changes (Bird 1988). The upper limit of *Galaeolaria*, which stands between mid-tide and mean neap tide level (mean spring tide range is here about 0.9 m), rose about 1 cm between 1986 and 1996, while the lower limit has been much modified by black mussel (*Mytilus* spp.) predation (E.C.F. Bird, personal communication, 1996; photo E.C.F. Bird, Feb. 1996)

Galeolaria caespitosa forms ledges and cushions on rocky shores and artificial structures such as sea walls and piers (Fig. 39), and could be used as an indicator of sea-level change (Bird 1988, 1993).

Shells not in growth position usually provide less reliable indications of sea level than *in situ* material because reworked shells may have been left at any level in the intertidal range, or even swept well above high-tide level by storm waves. On the sea floor, shells can undergo landward transport to misleading levels (Macintyre et al. 1978). Similar remarks can be made about most of the material cored from the floor of the continental shelf, where it is usually difficult to distinguish what is *in situ* from what is not. According to Petersen (1986), there are few reliable sea-level indicators in the group of molluscan species and they are all exposed to *post mortem* transport. However, on the basis of quantitative studies of subfossil molluscan assemblages, it may become possible to differentiate between littoral, shallow- and deep-water fossil environments. In the same way, by the use of detailed micropalaeontological analyses of ostracods, dinoflagellates and foraminifera, palaeowater depths and other palaeoenvironmental factors can, in some cases, be assessed (Wingfield 1995).

2.2 SEDIMENTARY SHORES

Unconsolidated marine sediments such as mud, sand, pebbles, or shell and coral fragments can be deposited in many coastal environments. The approximate sea level at the time of their deposition may be inferred, in many cases, from granulometric, (bio)stratigraphic, and physico-chemical evidence. The grain size of a marine deposit depends largely on the energy of the water environment, which can vary, at the same water depth, with exposure to waves and currents. Marine muds and clays are deposited only in very calm water, either at depth (below the reach of wave movement), or in very sheltered coastal basins (mudflats in lagoons and estuaries). Flora or fauna still in position of growth, and other parameters (e.g. local tidal range and topographic configuration), will often help to indicate the kind of environment in which they formed and the depth of their deposition, and to estimate adequate error margins. For example, former sea level has been determined to within ±10 cm from coastal palaeomarsh deposits, even in macrotidal areas, by using certain foraminiferal assemblages (Scott and Medioli 1986). Botanical macroremains can also contribute in various ways to sea-level reconstructions, by determining the degree of salinity and therefore the position relative to the local mean high-water level (Behre 1986) or the influence of former fresh-water inputs. Assemblages of fresh-water, brackish and marine microflora (diatoms) can be used to infer marine transgressions or regressions (Palmer and Abbott 1986).

In glacio-isostatically uplifted areas, the highest trace of marine action preserved on a coast is called the "marine limit". This can be identified, according to Andrews (1986), either as the lower limit of perched glacial boulders, till and continuous terrestrial deposition, or as the upper limit of wavewashed bedrock, shore deposits, beach ridges and *in situ* marine fossils. In ice-contact situations, the maximum recorded sea-level elevation usually corresponds with flat and channelled surfaces on massive glacio-marine deltas, where former low-tide position can be identified at the elevation where channels disappear from the surface of the delta.

In tropical regions, mangrove is probably the best sea-level indicator among intertidal vegetation. It extends into subtropical areas as far north as St George's Parish (32°23'N) in Bermuda, and as far south as Corner Inlet (38°45'S) in Australia. Mangrove forests are best developed where there is an extensive suitable intertidal zone (as found on low-gradient or macrotidal coasts), and an abundant supply of fine-grained sediment. They are more luxuriant in areas of high rainfall or abundant fresh-water supply through run-off or river discharge. They may also grow, however, on sand, peat or coral substrates (Woodroffe 1990, 1992).

The genera *Rhizophora* (Fig. 40) and *Avicennia* (Fig. 41) are most common, though there are many others, especially along Indo-Pacific coasts. The impacts of sea-level changes on mangroves have been

Figure 40   The roots of *Rhizophora* mangrove form a tangle of many arches above the intertidal mud, making access very difficult. Saloum River estuary, Sénégal (Photo 3031, Dec. 1976)

Figure 41   The roots of *Avicennia* mangrove trees, called pneumatophores, project from the intertidal mud and are used by the plant for respiration. Among the living roots, several dead *Porites* microatolls, capped by mud, are visible. The corals date from about 6 ka BP and indicate the position of low tide level at that time. Near Maudulung, Sumba Island, Indonesia. Local tidal range is 2.8 m. The water level is near MSL (Photo D299, Aug. 1991)

investigated in several studies (Woodroffe 1988, 1990; Ellison 1989; Woodroffe and Grindrod 1991; Ellison and Stoddart 1991; Bird 1993) and there is general agreement that fossil mangrove deposits, especially above present high-tide level, can be used to indicate former sea levels.

Peat layers are useful in stratigraphical reconstructions of sea-level histories. Fresh-water peats can be formed by stagnant water at any altitude, and only indicate, therefore, that the sea was at a lower level than that at which the peat formed. When intercalated with marine layers and dated by radiocarbon, peat can provide maximum or minimum age estimates for the marine transgressions. Brackish peats will of course give more precise indications on the sea-level proximity. The elevation above sea level at which a brackish peat may form depends on the topography, ground permeability, tidal range, and fresh-water flow. Brackish peats can therefore develop at any level in the intertidal range, though the most common level of formation is probably near MSL (Jelgersma et al. 1975).

After their deposition, the levels of a peat layer are often affected by autocompaction (Kaye and Barghoorn 1964), caused by decreased permeability with density changes, plastic deformation, loss in volume with decomposition, and complex physico-chemical changes that have a continuing effect on the structure and strength of the peat fabric. The survival of a salt marsh capping a peat formation will depend on the balance between the rate of autocompaction and that of new sedimentation. The impact of any change in sea level on the surface salt marsh will therefore be equivalent to that of an increase or a decrease in the autocompaction rate. Peat formation in the coastal zone is quite possible under the influence of a rising sea level as long as vertical accretion of sediment is slightly greater than the corresponding rate of sea-level rise plus lowering rate due to autocompaction. Consequently, the formation of peat layers intercalated with clastic layers is more likely during a relatively slow sea-level rise. During such a rise in the water table there may have been minor oscillations of sea level, which are masked by the peat growth, but a significant lowering of the water table would have caused an interruption in the peat growth (Streif 1979–1980).

Owing to compaction, which may reach as much as 90% of the initial thickness (Wiggers 1954), the use of peat layers for sea-level reconstructions has large uncertainty margins. For this reason, samples from the base of the peat bed, especially where it rests on much less compactable deposits, are preferred to surface- or middle-layer samples as sea-level indicators (Jelgersma 1961; Van de Plassche 1982).

Sand deposits are common where wave energy is moderate to large, and are found on beaches and intertidal shores and near lagoon entrances. Sand is also found in dunes and river deposits, and sedimentological evidence is needed to avoid confusion between these depositional environments. Lastly, pebbles are generally left only in high-energy shores, usually on exposed beaches, with pebble sizes sometimes increasing on the upper beach.

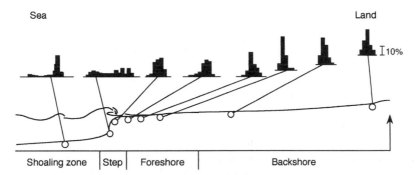

Figure 42 Grain-size histogram variations across a beach profile west of Sidi-Ferruch, Algeria (after Degiovanni, 1973)

The above criteria are quite general, however, and sequences of different morphological units, either permanent, seasonal or even episodic, corresponding to the same sea level may be found side by side in the same coastal section; their correct interpretation is not always an easy task.

Of special interest as a sea-level indicator in a beach profile is the step zone where the waves break. On the shores of microtidal seas, like the Mediterranean, this step is located in the zone where the swash and the backswash meet, and its top is normally covered by only a few centimetres of water. Here, grain-size analysis may show polymodal histograms, in contrast to the unimodal ones of the backshore and the bimodal ones of the nearby shoaling zone (Fig. 42). This can be explained by typical dynamics patterns in the step zone (Miller and Zeigler 1958). In fossil counterparts, according to Dabrio et al. (1991, p. 387), the step may be preserved in two ways, depending on the predominant grain size: (a) in gravelly beaches it is marked by an accumulation of the coarsest grain sizes available; and (b) in sandy beaches it is marked by a change from parallel lamination (which is common in the foreshore swash zone) to tabular bedding. Such a grain-size sequence may be found especially in microtidal beaches which receive sediments from the sea and the land, but is more rare where the source of sediments is unique (as, for example, on barrier beaches).

Several grain-size indices have been used to classify beach sediments; they are generally descriptions of grain-size distribution. Their interpretation remains controversial, because the relationship between such indices and transport and deposition processes is at best empirical and difficult to demonstrate.

The term "beachrock" applies to beach sediments deposited within the intertidal zone (often in the upper part of it) and cemented by calcium carbonate. It is generally limited to warm waters (in the range of coral reefs) or to some areas of temperate waters (e.g. in the Mediterranean,

Figure 43   Slightly emerged beachrock near Kyrenia (Cyprus) (Photo A831)

Fig. 43). Beach materials may have a great range in size, from fine sands to large boulders, but are generally very similar to the materials of the adjacent shore. Intertidal cementation of beach deposits has been ascribed to various causes: microbial action, inorganic precipitation from ground water, and inorganic precipitation from sea water (for reviews, see Kaye 1959; Guilcher 1961; Russell and McIntyre 1965; Hopley 1982; Dalongeville 1984; Bernier et al. 1990; Nunn 1994). Some evidence suggests that the cause of beachrock lithification may not be the same everywhere. Most authors nevertheless recognize that beachrock forms within a beach, and has been exposed in its present position on the beach by erosion (Hopley 1986a). The thickness of the cemented strata varies with the tidal range. Thin in tideless areas, the beachrock may reach over 3 m in thickness in macrotidal environments. Its upper surface usually shows a superimposition of beachrock slabs, each showing the same seaward slope as the nearby beach, delimited landward by basset edges.

   A beachrock is generally a good indicator of sea level (with a vertical uncertainty depending on the local tidal range). Radiometric ages of beachrock samples (shells, coral debris) should, however, be interpreted with caution, because these may be reworked deposits, older than the beach formation; in addition, beach deposition must necessarily precede the lithification of the beachrock. Apparent ages of organic beach materials would provide only maximum dates for the sea level at which lithification occurred; a date obtained from the cementing matrix could indicate the age of lithification, but great difficulties may arise in trying to obtain such cement without also including fine-grained carbonate beach materials.

Diagenetic products corresponding to changes in degree of water saturation of sediment pore space, rate of pore fluid movement, fluid energy and chemistry, and a variety of other factors, may also provide evidence of former sea-level positions. These diagenetic products have been discussed by Coudray and Montaggioni (1986). They include polygonally arranged desiccation mudcracks, caused by shrinkage of carbonate muds associated with the marine supratidal environment; primary holes (*fenestrae*) and sediment detritus (*intraclasts*) in the rock framework, which are indicative of upper intertidal and supratidal deposition; the occurrence in carbonate rocks of internal sediments, which imply near-surface conditions, or of secondary pores, which can develop only in fresh water; lastly, a wide range of cementation processes, some of which correspond closely with a specific marine zonation. For example, intergranular cements of beachrocks are typically *vadose* (i.e. corresponding to the intertidal or supratidal zone) and those of a reef framework are typically *marine phreatic* (i.e. subtidal).

Such criteria can be especially useful in determining former sea-level positions in reef areas. On a reef flat, for example, accumulation of skeletal detritus, torn away from the outer reef by storm waves and filling in depressions and irregularities in the reef framework, often commences in shallow water. Here, lithification starts in the subtidal zone, the intergranular spaces being gradually occluded with marine phreatic cements. In the intertidal zone, intergranular cements are mainly of marine vadose types. In the emerged coral conglomerates, the method of investigation consists of sampling vertical sections at several levels and identifying, by petrological analysis, the former environment (marine phreatic or vadose) of the first generation of intergranular cements in each sample (Fig. 44). Once the boundary level between the two former environments of cementation has been found on the vertical section, the amount of the possible sea-level change can be estimated from the difference in level between this boundary and present mean low-tide level (Montaggioni and Pirazzoli 1984; Pirazzoli and Montaggioni 1988).

On the continental shelves, seismic stratigraphy gives a picture of the way the rock strata have been piled on one another. Each rise and fall of the sea removes some sediment from continental margins and leaves piles of new sediment, producing recognizable patterns such as erosional unconformities. These features can be used to identify the best places to drill the rock for samples and to interpret sea-level changes. By applying this technique, various sea-level curves have been proposed, the most complete probably being that proposed by Haq et al. (1987) to summarize global sea-level changes in the last 250 Ma. For the last 20 ka, which corresponds to the last deglacial hemicycle, high-resolution seismic

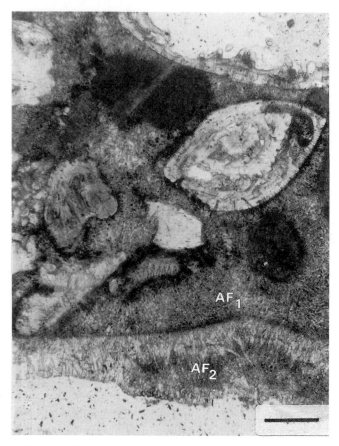

Figure 44   Coral conglomerate sample showing evidence of a relative sea-level fall
(phreatic, then vadose cementation). The earlier stage has densely packed aragonite
needles (AF1) with chisel-shaped ends, while the later stage includes regular,
aragonite fringes of blunted fibres (AF2). The scale is 0.1 mm. From a sample
collected from the Bora Bora barrier reef (French Polynesia) 0.4 m above MSL
(analysis and photo by L.F. Montaggioni)

reflection profiles can provide useful preliminary information to identify
submerged shorelines, to be completed with more detailed work.

Rough estimation of sea-level changes can also be obtained, with assumptions, using proxy data, such as changes in the oxygen isotope record. The
oxygen isotope ratios, which are obtained by analysing foraminifera in deep-
sea sediment cores, can give an approximate estimation for the last few
million years of global ice volume (and therefore of global sea-level changes),
because the isotopic composition of an ice sheet is generally lighter than that
of the ocean water. This means that a change of about 10 m in the global sea
level is equivalent to about 0.1‰ change in the isotopic record. As this is

close to the present lower limit of resolution in isotopic measurements, this method can only indicate sea level within ±10 m, when other sources of uncertainty (such as variations in ocean temperature, variations in sedimentation rates, bioturbation) are disregarded. The chronology from ocean cores has been calibrated with geomagnetic reversals, biostratigraphy, and comparison with astronomic periodicities.

Global sea-level changes can also be interpreted from the study of ancient shorelines that have been tectonically uplifted or submerged, if the rate of uplift or subsidence can be assumed to have remained constant in time. In coral reef areas, in particular, the reef itself may be considered to be like a continuous tape recorder, each reefal element developing when the rising sea overtakes the land. Thus, in uplifting areas, reef crests represent the peaks of each transgression. In a similar way, narrowly zoned corals can be used to reconstruct a continuous regional sea-level curve if uplift rates are known.

2.3   ARCHAEOLOGICAL AND HISTORICAL SEA-LEVEL INDICATORS

2.3.1   Archaeological remains

Most archaeological remains give no evidence on how far from sea level they were formed. In only a few cases is it possible to specify whether the relative sea level has risen or dropped near ancient sites since the time they were inhabited. A site on dry land, the elevation of which has increased, is usually hard to distinguish from a site which was constructed at a higher elevation or further inland. In certain cases, however, the occurrence of elevated marine features on the remains, or elevation inconsistencies in artefacts which had clearly been intended for use close to sea level, may reveal that there has been a change in relative sea level.

Submerged remains are generally evidence that sea level has risen, or the land level subsided. However, confusion should be avoided with artefacts which were formed underwater, or in the tidal zone. The latter, if identified, are obviously most useful, because they imply specific activities closely related to the sea (such as sailing, fishing, boat-building, salt production, shell gathering). Two main categories of archaeological remains related to the sea can be distinguished: (1) those which must have been located near a shoreline and, according to their use, must have remained above or below the sea and (2) those belonging to structures partly underwater and depending for their use on tidal fluctuations and marine conditions (Flemming 1979–1980).

Precise indications can be provided by remains of the second category, such as slipways (Fig. 45), certain harbour constructions (Paskoff et al. 1985), and especially fish tanks (Pirazzoli 1976, 1988), from which relative sea-level change can be estimated with accuracy.

Figure 45   This slipway cut into the limestone rock in Antikythira Island (Greece), was suddenly made worthless by a 2.7 m uplift, probably in AD 365 (Photo 5040, Sep. 1979)

Structures which have been built with their foundations on dry land, but are now partly or totally submerged, can only provide minimum estimations of relative sea-level change, since their original elevation above sea level is generally unknown. The latter category includes submerged remains of houses, churches and temples, warehouses and shops, mosaic floors, tombs, flights of steps, tunnels, passageways, doorways, storage tanks, silos and bins, quarries (Fig. 46), tells, middens, and drains and gutters (Flemming 1979–1980). Prehistoric submerged remains may also include dolmens, shell mounds, graffiti or paintings on walls of sea caves, as well as *in situ* stratigraphic evidence (Masters and Flemming 1983; Bailey and Parkington 1988).

2.3.2   Historical data

Historical accounts may provide useful information on sea-level changes only for the latest part of the last 20 ka, covering periods which can last

Figure 46   A slightly submerged quarry floor south of the town of Rhodes, Rhodes
Island, Greece (Photo 4477, Apr. 1978)

2000 to 3000 years in the best cases, but only one or two centuries in many
other areas.

Historical accounts do not correspond generally with modern criteria of
scientific observation. They often need, therefore, to be reinterpreted, taking
into account theories, interpretations and methods of measurement that
were in use at the time. This reinterpretation can obviously provide only
qualitative information about islands which emerged or were submerged at a
certain time, about cities that were "drowned by the sea", about passages
which could be crossed on dry land (and are now submerged), or only by
boat (and are now located inland from the coast). In these accounts, the
effects of relative sea-level change are not always easy to separate from those
of possible sedimentation, erosion, landslides or tsunamis.

Certain descriptions made by ancient writers are essential to understanding how certain constructions were related to sea level. For example,
we learn from Varro (116–27 BC) how marine fish tanks were used by rich
Romans, following a fashion that started during his time; Columella
(1st century AD) gives details of their construction (Fig. 47) and their
relationships to tidal levels; lastly, Pliny (AD 23–79) confirms that the
fashion for fish tanks must have seen a rapid boom throughout the
Mediterranean, and mentions them in places as far distant as Narbonne
and Phoenicia, with many on the coasts of the Tyrrhenian Sea, near Rome
(Pirazzoli 1988). Many remains of Roman fish tanks, all dating from

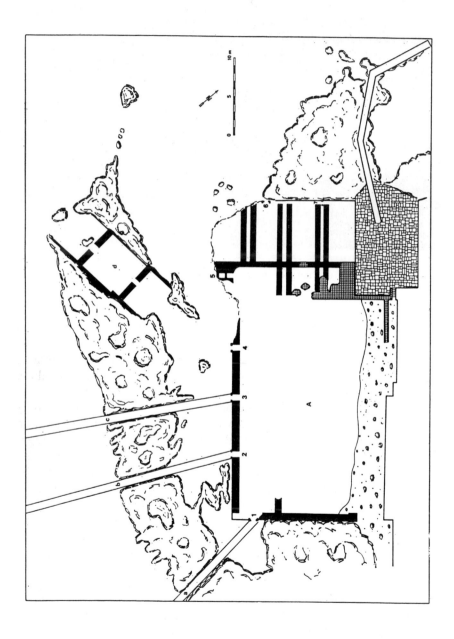

Figure 48   A submerged fish tank cut into rock at Mokhlos, Crete, Greece. It consists of two rectangular compartments divided by a wall of rock. Two channels leading from the compartments seawards are visible. A third channel, less visible, exists on the right (Photo 4846, May 1978)

between the 1st century BC and the 1st century AD still exist throughout the Mediterranean (Fig. 48). Their study has enabled former sea-level positions to be determined in certain areas. Along the coasts of the Tyrrhenian Sea, for example, such data demonstrate that 2000 years ago the sea level was about 0.5 m below the present level, but was rising by about 0.75 mm/yr between 50 BC and AD 150 (Pirazzoli 1976).

Certain coastal descriptions made by ancient Greek or Roman historians or geographers have proved useful in estimating changes that have subsequently occurred, especially when distances between localities were mentioned. In many cases, however, subsequent geomorphological changes may have been on such a scale that field coring was necessary to find the former shoreline at ancient sites such as Marathon (Baeteman 1985) or the Thermopylae (Kraft et al. 1987).

Over a short period, ancient coastal maps can provide useful information for coastline advance or retreat when compared with modern maps, with air photographs or satellite images, though past sea levels can hardly be

Figure 47   (*opposite*) The Roman fish tank at Torre Valdaliga (Civitavecchia, Italy), now partly submerged, consists of a main rectangular basin (39 m×19 m) with inner subdivisions (A), and a smaller basin cut into an abrasion platform (d). The basin A was connected to the sea by three supply channels (a, b, c), 27 to 31 m long, also cut into the platform. Vertical grooves at the basin entrances (1 to 6) were used to hold bronze grids with a fine enough mesh to prevent the fish from escaping, while permitting water exchange with the sea (after Schmiedt 1972)

determined from such evidence. Lastly, ancient plans of bridges (Fig. 49) or other buildings, or even certain paintings can be of use for comparisons with the present situation.

2.4 DATING A SEA-LEVEL INDICATOR

Apart from historical and archaeological dating, age estimations of sea-level indicators are generally based on radiometric methods applied to field samples. For the last 20 ka the most commonly used method consists of measuring the radiocarbon left in organic samples. Carbon 14 (^{14}C) is continuously produced by the interaction of cosmic ray neutrons with nitrogen atoms in the atmosphere, and the dating method is based on the fact that living plants and animals incorporate ^{14}C with the same concentration as in the atmosphere. After their death, ^{14}C disappears from their tissues by radioactive decay at a known rate, so that (with some assumptions) the time elapsed after their death can be determined by measuring ^{14}C left in the sample at present.

The *conventional ^{14}C age* reported by all laboratories is defined by using the following assumptions and internationally agreed conventions (Mook and Van de Plassche 1986):

(1) the ^{14}C activity is defined internationally (as standard oxalic acid distributed by the US National Bureau of Standards) and is assumed to have always been the same (in fact, as shown below, some natural variations existed, which have caused discrepancies);
(2) there has been no inclusion of new ^{14}C in the sample after its death;
(3) the ^{14}C half-life ($T_{1/2}$) used is 5568 years (in fact this is known to be in error by 3%; the better value of 5730 years should not be used, however, in order to avoid confusion with earlier reported dates);
(4) ^{14}C ages are given in years before present (BP), i.e. before AD 1950, and have to be corrected for isotope fractionation, based on the measured $^{13}C/^{12}C$ ratio.

An age measured by radiocarbon (T) is always given with the standard deviation ($\pm\sigma$), which represents the statistical uncertainty of the measurement. With the assumptions used, there is a 68% chance that the true age of the sample is included between $T + \sigma$ and $T - \sigma$, a 95% chance that it is between $T + 2\sigma$ and $T - 2\sigma$, and a 99.7% chance that it is between $T + 3\sigma$ and $T - 3\sigma$.

When comparison with archaeological or historical dates is necessary, conventional ^{14}C ages are calibrated into astronomic ages, using conversion tables, diagrams or computer programs (e.g. Stuiver and Reimer 1986; Stuiver and Braziunas 1993).

The careful choice of samples to be dated is essential in order to obtain reliable age estimates. Recrystallization and isotopic exchanges between the

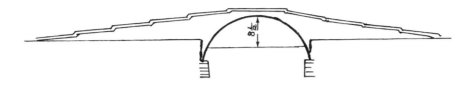

Figure 49   The present-day Paglia Bridge, near the Doge's Palace in Venice, Italy, was built in AD 1360. A 16th century drawing of the bridge, preserved at the Archivio di Stato dei Frari, Venice (*Raccolta disegni Rialto: Ponte de la Pagia*), from which the sketch is derived, indicates a height of 8.5 Venetian feet (2.76 m) above the "chomun de l'aqua". The datum of that time, the so-called "comune marino", was the horizontal line topping the dark-green band of algae at the borders of the canals, approximating local MHW level. In 1973, the corresponding height was about 2.30 m, indicating that a relative sea-level rise of 0.46 m has occurred (Pirazzoli 1974). The photograph shows the present-day situation of the bridge; the horizontal line of the datum is clearly visible (Photo E784, Jul. 1996)

sample material and the environment will modify the apparent age of a sample and give misleading results. Samples that may have been contaminated by older or younger organic material (such as the roots of trees or residues from burrowing animals), or carbonate, should also be carefully avoided. Various cases of possible contamination, especially in peat samples, have been discussed in detail by Mook and Van de Plassche (1986).

Carbon 14 measurements of dendrochronologically dated tree rings have shown that assumption (1) above is not entirely true, because variations in the ^{14}C activity of atmospheric CO_2 have occurred. Small variations (3‰ or less) are correlated with the 11 year sunspot cycle. Other variations, ascribed to changes in solar activity, have produced changes in ^{14}C ages of a few hundred years within a historical period of less than half a century. Finally, a long-term trend exists, for ages older than 2000 years, which causes radiocarbon ages to appear systematically younger than they should be. Dendrochronological dates for the last 13 ka have shown the reality of this systematic trend, which has also been confirmed back at least 30 ka using a comparison with U/Th ages (Bard et al. 1990b).

Uranium/thorium dating, when possible, should therefore be preferred to ^{14}C dating, especially for Late Pleistocene marine samples, because it does not require independent calibration (Edwards 1995). Unfortunately, the materials that can give reliable results with the U/Th method, and the number of U/Th dating laboratories available, are much less than for the ^{14}C method, and in addition the cost of U/Th dating is higher than that of ^{14}C dating.

Other dating methods useful in the study of sea-level changes during the last 20 ka include varve counting and palynological age estimations.

2.5   HOW SEA-LEVEL CURVES ARE CONSTRUCTED

To be plotted on a bidimensional graph, a sea-level index point should be summarized by at least two numerical values: elevation (or depth) of the corresponding sea level, and age. To be comparable with other sea-level index points, uncertainty ranges arising from the sampling and dating techniques employed, and taking into account possible errors resulting from interpretation, should be specified for each value. When several dates from a tectonically homogeneous area are available, they should be graphed as a series of error rectangles, error crossed segments, error ellipses or error bands (Fig. 50). These uncertainty margins are essential for interpretation, and should accompany any linear sea-level curve presented.

Figure 50   (*opposite*) Example of the construction in three stages of a reliable sea-level curve for the Fenland (eastern England): (a) initial plot of MHWST index points with error margins; (b) possible MHWST error band deduced from the preceding data set (Shennan 1982a); (c) updated error band of MHWST deduced from more complete data (Shennan 1986)

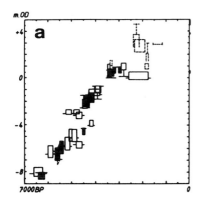

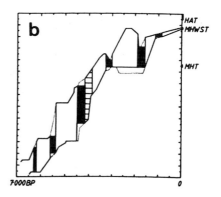

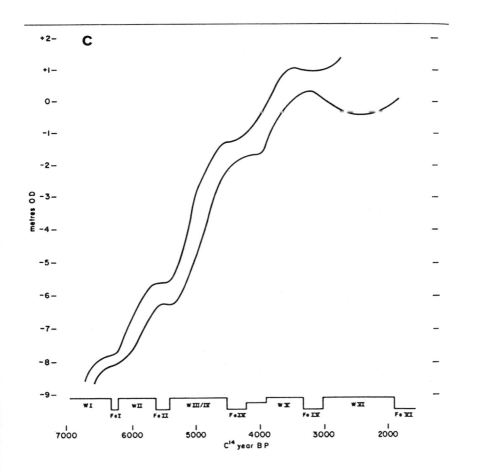

A sea-level curve alone is a poor summary of sea-level changes if the variation in data points is not shown. This is because a single curve may reflect not only the sea-level data, but also subjective ideas and in some cases preconceived theories; as the interpretation of changes in sea level is scale-dependent (Shennan and Tooley 1987), users of the information should be able to use original sea-level index points rather then the linear curve, which in addition may be an inappropriate summary for their analysis. Unfortunately, many published sea-level curves are just continuous lines fitted through sea-level index points, with little regard to the errors likely to have affected them; this represents a regrettable loss of information in the available literature, and sometimes an unnecessary source of misinterpretation and misunderstanding. Only when more complete information is provided on the sea-level index points used can linear schematizations become acceptable for use in comparisons with data from the same and other areas.

Chapter Three

The ice age Earth

3.1 HOW THE LAST GLACIATION DEVELOPED

Recent advances in Quaternary studies have greatly improved our knowledge of Earth history during the last climatic cycle. The study of ice cores drilled by an American team at Camp Century, then at Dye3 (Greenland), by a Russian team at Vostok (Antarctica), and by European (GRIP) and American (GISP2) teams in central Greenland, have produced much data on the natural variability of climate (Jouzel 1994). Oxygen isotope analyses of foraminifera from deep-sea sediment cores have contributed to estimates of changing global ice volumes and consequent glacio-eustatic sea-level changes (see Section 1.2).

After the last interglacial period (*c.* 140–110 ka BP), which seems to have been warmer than the Holocene (up to 4°C warmer at GRIP, a little less at Vostok), a gradual cooling interspersed with minor climatic oscillations occurred during the following 90 ka. Glacial–interglacial sequences show an approximate 100 ka periodicity, which has been ascribed to the 100 ka cycle of orbital eccentricity. The periodicity in glacial–interglacial recurrences would result from the variability of summer insolation at about 65°N and from a non-linear response of ice sheets to orbital forcing (Lorius et al. 1985).

As a result, after the end of the last interglacial period, new ice domes developed gradually over northern America, northern Europe, and probably the Barents and Kara Seas, in addition to the "permanent" Antarctica and Greenland ice sheets, whereas minor mountain ice sheets thickened in the Andes, the Alps and in some other upland areas. Some of these ice domes might have existed only for a relatively short time (e.g. between 22 and 14 ka for the Barents Sea ice sheet, according to Hebbein et al. 1994).

High-resolution isotopic analysis of ice cores from Greenland reveals additional irregular but well-defined episodes of relatively mild climate during the middle and late parts of the last glaciation. These Greenland interstadials, some of which are very attenuated or absent in the Antarctic ice cores, began abruptly, perhaps within a few decades, but terminated gradually or in a stepwise fashion. They sometimes occurred at irregular intervals (Fig. 51). The duration of these interstadials ranges from about 500 to 2000 years, and their irregular occurrence suggests complexity in the behaviour of the North Atlantic Ocean circulation. Johnsen et al. (1992) suggest that such glacial interstadials were linked to changing circulation in an ocean partly covered by sea ice.

Towards the end of the last glaciation, abrupt climate changes (within a few decades) of at least regional extent seem to have occurred in the North Atlantic region (Dansgaard et al. 1993). This contrasts with the relative stability observed in the same area during the Holocene.

3.2   THE SEA-LEVEL POSITIONS DURING THE LAST ICE AGE

A rough approximation of changing global ice volume (and therefore of global sea-level changes) can be obtained from the 18 ka BP perimeter of continental ice sheets, which is reasonably well-known, at least for the Laurentide and Scandinavian ice sheets, and on assumptions concerning ice mechanics of equilibrium glaciers. Estimates for ice-volume sea-level equivalents range from as high as 163 m to as low as 102 m (Matthews 1990).

Another kind of estimation can be deduced from the oxygen isotope record which is obtained by analysing foraminifera in deep-sea sediment cores. This method is based on the fact that the isotopic composition of an ice sheet is generally lighter than that of the ocean water. Shackleton and Opdyke (1973) suggested that a change of about 10 m in the global sea level corresponded with a 0.1‰ change in the isotopic record. Fairbanks and Matthews (1978) obtained a field calibration of 0.11‰ for a change of 10 m by measuring the isotopic composition of corals from raised terraces in Barbados. These can be only approximations, however, and Shackleton (1987) has summarized various sources of uncertainty, for example variations in the temperature of the ocean water, bioturbation, variations of the average isotopic composition of the former ice sheets in their size and their latitudinal position. The accuracy of sea-level estimation from oxygen isotope records depends therefore on the validity of the assumptions made and could correspond, at best, to the precision of $\delta^{18}O$ measurements, i.e.

Figure 51   (*opposite*) Comparison, over the last 50 ka, of the climatic records, from Vostok, Antarctica (deuterium content), from GRIP, Greenland (^{18}O content), and from the North Atlantic site V23-81 (percentage of *N. pachyderma*, an indicator of sea-surface temperature) (from Jouzel et al. 1994)

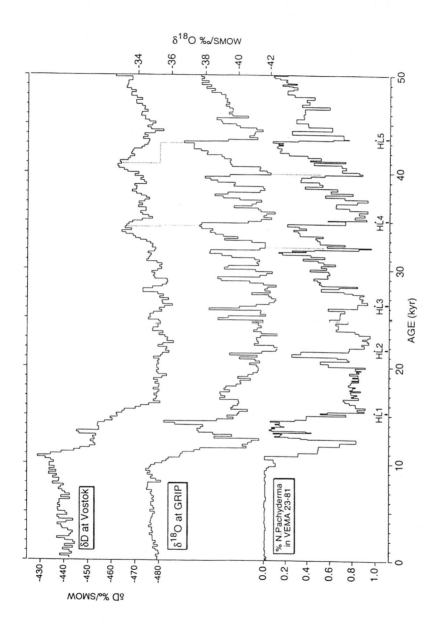

to 0.1‰, or to ±10 m in equivalent sea level. With these assumptions, the global sea-level rise since the last glacial maximum can be estimated to be about 130 m (Chappell and Shackleton 1986).

Suggestions on local sea-level positions during the last glacial maximum are often reported in the literature. This is of course not the case in formerly glaciated areas, where marks of former shorelines on ice margins have disappeared with ice melting. Along the margins of continents or islands remote from ice sheets, remnants of former shorelines still exist in many areas, submerged by the subsequent eustatic sea-level rise but not yet capped by younger sediments that would make their identification uncertain. There is much bathymetric evidence suggesting a sea stand around −90 to −130 m or even more, but it is most often undated or it has not been satisfactorily shown to belong to the 18 ka BP shoreline. In only a very few cases can former sea-level position be determined in the field with a reasonable range of vertical uncertainty margins and dated with accuracy.

Curray (1961) dredged shells of predominantly shallow-water organisms from a depth of about 119 m near the edge of the continental shelf south of Mazatlan (Mexico), which were radiocarbon dated to 19 300 ±300 years BP. In a subsequent compilation, Curray (1965, p. 724) admits that the "depth of the low stand is not well established, . . . but it appears that the maximum lowering was at least 110 m". His estimate is of 124 m.

In the Timor Sea, a littoral mollusc shell, *Chlamys senatorius*, collected from about −130 m, was dated to 16 910 ±500 years BP (Van Andel and Veevers 1967).

Milliman and Emery (1968) inferred a minimum sea level around 15 ka BP at about −125 m from shallow-water material dredged or cored from a very wide area of the Atlantic continental shelf off the United States. However, this 15 ka BP date, later than the usually accepted date of 18 ka BP, was not confirmed by subsequent studies. Also, the frequent use of relict oyster shells and other shoreline deposits from the continental shelf as sea-level indicators was criticized by Macintyre et al. (1978), owing to evidence of significant postdepositional and landward transport of these shells.

In the Arafura Sea, a shallow-water fossiliferous algal rock, recovered by dredge from a submerged reef between −130 and −175 m, gave a radiocarbon date of 18 700 ±350 years BP. At some 380 km distance on the same shelf, a core struck a wood sample at about −100 m, which was radiocarbon dated to 14 500 ±700 years BP, suggesting that marine transgression at this level occurred after that time (Jongsma 1970).

Two *in situ* samples were collected with the mechanical arm of a submersible near a marine terrace at −160 to −165 m off One Tree Island (Great Barrier Reef, Australia). A large colony of aragonitic *Galaxea clavus* (Dana) (a shallow-water reef coral, which has also been observed at some depth), collected at −175 m, has been dated to 13 600 ±220 years BP by radiocarbon and to 17 000 ±1000 years BP (two dates) by ^{230}Th (Veeh and Veevers 1970). In the same area, a fossiliferous calcarenite identified as a beachrock, collected at −150 m, was radiocarbon dated to 13 860 ±220 years BP by the same authors.

On the continental shelf from Guinea to Sierra Leone, an algal rock broken from between −103 and −111 m contained a *Porites bernardi* coral which was radiocarbon dated to 18 750 ±350 years BP (McMaster et al. 1970).

Off south Barbados, Fairbanks (1989) estimated that sea level was at −121 ±5 m around 17 ka BP from drilling of *Acropora palmata* corals, assuming a continuous local uplift trend of 0.34 mm/yr.

Two samples from a reefal layer, collected with a land-based deviated drill at *c.* −170 m from the side of Mururoa Atoll (French Polynesia), have been dated by the U/Th method to 15 585 ±50 and 17 595 ±70 years BP (2σ) (both these dates are weighted means of three replicates). The same samples have been radiocarbon dated to 13 060 ±140 years BP and 14 735 ±150 years BP (2σ), respectively (both dates are weighted means of five age determinations) (Bard et al. 1993).

An *Acropora* coral collected in 1991 from the outer reef slope of Mayotte Island, using a small two-seater submersible, at a depth of approximately −152 m, was dated by U/Th 18 400 ±500 years BP; it is considered to correspond to the lowest relative sea level of the last glacial maximum, which seems to have lasted no more than 500 years (Colonna 1994).

Lastly, on the continental shelf off western Sardinia, apparently *in situ* vermetids dredged from a submerged cliff between −126 and −150 m have been radiocarbon dated (without reservoir correction) to 18 860 ±170 years BP (laboratory reference GifA 95431), whereas *Dendropoma* vermetids have been dated to 12 300 ±80 years BP (R-2264, collected between −140 and −155 m), 12 910 ±120 years BP (GifA-95430, between −117 and −120 m), 13 130 ±110 years BP (GifA-95429, between −130 and −150 m), and 14 090 ±130 years BP (GifA-95432, between −156 and −160 m) (S. Carboni, University of Cagliari and M. Arnold, CNRS-CEA, Gif-sur-Yvette, personal communications, 1995) (Fig. 52).

The above differences in level do not necessarily correspond to changes in the eustatic sea level, which are likely to have been limited to between 17 and 20 ka BP. They are more probably due to differing isostatic responses of continental shelf areas with varying depths, widths and geology, to deglaciation unloading or meltwater load. According to Walcott (1972) and Cathles (1975, p. 142), the observed sea-level change is expected to be as much as 30% larger than the actual eustatic change in mid-oceanic islands, since the density of the upper mantle is about 3.3 g cm^{-3}. As to the differences in age between radiocarbon and U/Th dates, Bard et al. (1990b) have shown that the former are systematically younger than the latter, which are more consistent with dendrochronological calibration, with a

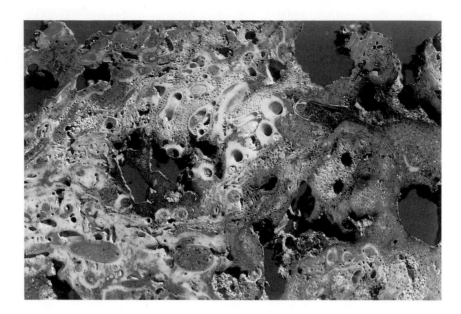

Figure 52   Detail of the vermetid shells dated 18 860 ±170 years BP on the
continental shelf off western Sardinia. As well-preserved shell material in sample
DR16 was insufficient to obtain a classical ^{14}C date, radiocarbon dating was
performed with the accelerator mass spectrometry (AMS) method by M. Arnold
(CNRS-CEA, Gif-sur-Yvette) (photograph taken by S. Carboni, University of
Cagliari)

maximum difference of about 3500 years at a radiocarbon date of *c.* 20 000
years BP. In other words, the maximum of the last glaciation, dated to *c.*
18 000 years BP by radiocarbon, may have occurred *c.* 21 500 sidereal years
ago.

3.3   LOW-SEA-LEVEL LAND BRIDGES AND LANDSCAPES

A recent useful interdisciplinary account of the geology and climate during
the last ice age has been provided by Dawson (1992). The reader may refer
to that work for general problems. Only a few points having a close
connection with sea level will be developed here.

Many continental shelf areas which are now submerged were exposed
land at the last glacial maximum (Fig. 53 (Plate I), Fig. 54 (Plate II)).
Among many land bridges existing with lower sea levels, the most
important for its effects on mankind and other mammalian migrations was
indeed Beringia, the great land connection between Asia and North
America. The formation of this land bridge in the Bering Strait also had

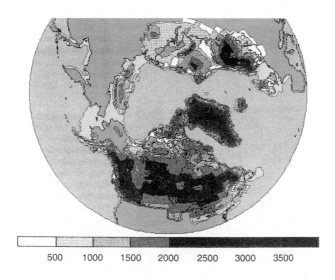

500 1000 1500 2000 2500 3000 3500

Plate I Figure 53 Regions that are now oceanic but were exposed land at glacial times (in yellow) according to Peltier (1994). The bar at the base of the map denotes ice thickness, in metres

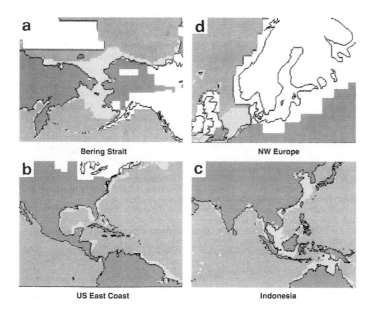

Plate II Figure 54 Detail of geographic regions that were exposed land at glacial times but which are now covered by ocean according to Peltier (1994): (a) Bering Strait; (b) US east coast; (c) southeast and east Asia; (d) northwest Europe

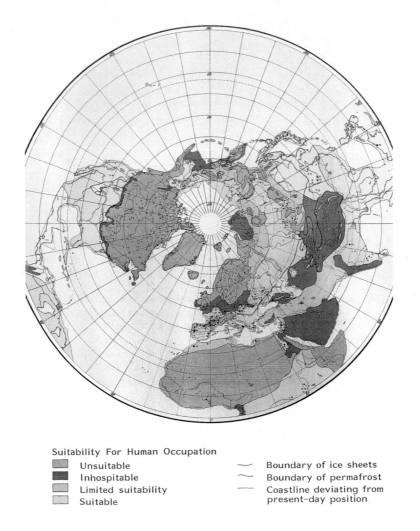

Suitability For Human Occupation

- ▨ Unsuitable
- ▨ Inhospitable
- ▨ Limited suitability
- ▨ Suitable

⌒ Boundary of ice sheets
〜 Boundary of permafrost
⌁ Coastline deviating from
present-day position

Plate III Figure 55 Human occupation of the northern hemisphere during the Upper
Pleniglacial of the Last Glaciation (about 24 to 15 ka BP), after Madeyska et al. (1992)

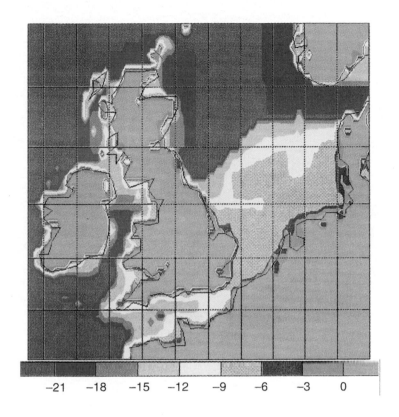

Plate IV Figure 56 Inundation map for the last glacial maximum land bridge that connected Britain to continental Europe. The bar at the base of the map denotes the time, in thousand years ago, that each geographical point on the land bridge was inundated by the ocean (after Peltier, 1994)

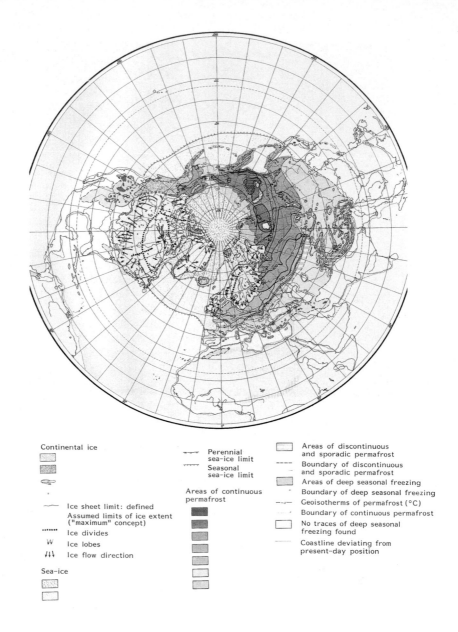

Continental ice

Ice sheet limit: defined
Assumed limits of ice extent
("maximum" concept)
Ice divides
W   Ice lobes
⧸↓↓ Ice flow direction

Sea-ice

Perennial
sea-ice limit
Seasonal
sea-ice limit

Areas of continuous
permafrost

Areas of discontinuous
and sporadic permafrost
Boundary of discontinuous
and sporadic permafrost
Areas of deep seasonal freezing
Boundary of deep seasonal freezing
Geoisotherms of permafrost (°C)
Boundary of continuous permafrost
No traces of deep seasonal
freezing found
Coastline deviating from
present-day position

Plate V Figure 62   Map of glaciation and permafrost in the northern hemisphere at the maximum cooling of the last glaciation (about 20 to 18 ka BP), after Conchon et al. (1992)

profound influence upon oceanic circulation, because it restricted water inflow into the Arctic Ocean from the area of the northern Pacific.

Sea level would have to fall only 46 m below its present position to expose a narrow land connection between the Chukot Peninsula and Alaska by way of St Lawrence Island. A reduction to −50 m would expose a second narrow connection north of the Bering Strait; and a reduction to −100 m would expose almost the entire area of the Bering–Chukchi continental platform (Hopkins 1967).

The Bering land bridge was probably in existence for most of the period between 80 and 14 ka ago. According to McManus and Creager (1984), in the Bering Strait area the relative sea level was lower than −68 m at 19 ka BP, near −55 m *c.* 16 ka BP and −30 m *c.* 12 ka BP.

In Fig. 55 (Plate III), only the Arctic continental shelf near the Bering Strait, exposed by the low sea level, and the Yukon River plain in Alaska are considered as "suitable for occupation" during glacial times, most of the North American continent being "unsuitable" as far as 50°N (Pacific coast) or even 40°N (Atlantic coast) because of the development of continental ice sheets or mountain glacial complexes.

Even at the glacial maximum, lowlands and many upland areas remained ice-free in Beringia. An immense area of continental shelf (the 200 m isobath to Bering Strait today is 700 km to the south and 900 km to the north) was covered by tundra and steppe vegetation and crossed by several rivers emptying into the Arctic Ocean (McManus et al. 1983). This land was an open route for migration of humans and mammals and might be expected to have been most attractive to human hunters (Hopkins 1983). Climate was drier than today, but grassland must have been more widely distributed, because palaeontologic evidence shows that large grazer mammals (horse, bison and mammoth) were common.

Having crossed the land bridge, the distribution of glacial ice would have been a critical factor affecting possibilities of further migration. Merged glacier systems extended almost continuously from the Arctic to the Pacific Ocean, blocking all land communication between Beringia and central North America. An ice-free corridor must have existed, however, during certain episodes of mild climate interstadials, when land communications were simultaneously possible between Siberia and Alaska and between Alaska and central North America. According to Hopkins (1967), this corridor was closed by ice earlier than 20 ka ago and remained closed until at least 14 ka ago and possibly until almost 10 ka ago, until the moment, or well after it, when the land bridge was closed again by the rising sea. According to Reeves (1983), however, environmental conditions would not have been sufficiently adverse to prohibit native occupation of two ice-free corridors, between the Cordilleran and Laurentide ice sheets, and along the coast.

According to genetic, linguistic and cultural differences between American Indians and Eskimos or Aleuts, Beringia must have seen at least two

waves of human migration. The precise time of the beginning of the
first wave is still uncertain although it can be estimated as at least 25 ka
ago, i.e. prior to the last glacial maximum. Older dates (aspartic acid
racemization ages of 45 ka for the Del Mar Man and of 70 ka for the
Sunnyvale skeleton) obtained from Californian sites are considered less
reliable (Bada and Finkel 1983). The first wave was that of the ancestors of
American Indians, part of which dispersed into central North America
along a temporarily ice-free corridor east of the Rocky Mountains. From
there, the way was open to the colonization of the rest of the continent,
east of the Andes and down to Patagonia, where their presence is attested
about 12.6 ka BP (Meighan 1983). Along the coasts of Chile, however,
evidence of Pleistocene human occupation seems to be missing (Paskoff
1970).

The second wave is dated from the early Holocene, when the earliest
known riverine and shoreline-dwelling Mongoloids reached the southwestern coast of Alaska around 8.4 ka ago (Hopkins 1967). As the land
bridge had long been drowned at that time, American Eskimos and Aleuts
may descend from groups that had crossed the Bering Strait by boat.

Another important area of land bridges for human and biotic migration
is that between southeast Asia, New Guinea and Australia. Most significant
in this region is the changing complex of islands and channels that
separated the continental shelves of Sunda and Sahul and conditioned the
movement of humans between southeast Asia and New Guinea–Australia.
Possible routes between the Sunda shelf and the Sahul shelf at lowstand of
sea level include, however, final water straits which by any route exceed
60 km. Such a distance implies some kind of navigation (Smith 1989).

The presence of humans is attested in the New Guinea Highlands and in
the Australian interior more than 25 ka ago, when the Sahul shelf was
exposed during the last glacial period and the two land masses were united.
Some evidence even points to at least two episodes of invasion of the
Australian continent: the first colonizers arrived more than 50 ka ago,
while a second wave followed 10 or 20 ka later.

The Sahul shelf region is considered by some authors to have been arid,
saline and inhospitable; however, according to Van Andel and Veevers
(1967, p. 108) the landscape of the Sahul Shelf "was attractive and suitable
for the migration of early man (seasonal herbaceous vegetation with open
forest along stream courses is probable)". Anyway, the Torres Strait, a
channel that narrowly separates Australia from New Guinea, with the
depth of the sill at −15 to −18 m, persisted as a land bridge for at least
60 ka, before being submerged in the Holocene.

The Torres Strait functions as a frontier zone between the hunter-
gatherers of aboriginal Australia and the horticulturalists of New Guinea.
During the glacial maximum, it formed an extensive upland plain, with
remnant karstic reefs and igneous hills offering restricted and diverse

habitats different from those of the rest of the shelf, which may have been important plant, animal and even human habitats, forming a chain along which contact was maintained between floristic and faunal elements north and south of the broad shelf (Barham and Harris 1983).

In northwestern Europe the floors of several present-day marginal-sea areas were exposed, though it does not seem that the Scandinavian and the Scottish ice sheets merged in a single mass. Great Britain was indeed connected to the European continent when the Channel and the southern part of the North Sea were emerged land (Fig. 56 (Plate IV)). According to Devoy (1985), no true land links are likely to have existed between Britain and Ireland during the maximum glaciation; from around 11.4 to 10.2 ka BP a discontinuous linkage formed by temporary islands may have existed between the north of Ireland and southwest Scotland, following the inter-action in time between isostatic change in land elevation and eustatic sea-level change. Nevertheless, a land bridge between Ireland and Scotland is indicated by Peltier (1994) in Fig. 56 (Plate IV), also at the time of the glacial maximum.

In the Mediterranean, a great part of the Adriatic Sea floor was exposed as a wide fluvio-lacustrine plain. The coastline was located at the northern edge of the Meso-Adriatic Depression, off Pescara (Colantoni et al. 1979). Almost all of the Dalmatian islands (over 1000 in number) were connected to the continent by the low sea level (Šegota 1982 1983). Remnants of the periglacial climate of that time are loess deposits found sporadically along the coast for about 1600 km, from the Quarnero islands to the Lake of Skadar (Markovic-Marjanovic 1971). It is typically developed in the outer Quarnero island of Susak, where four loess layers appear in superposition with fossil soils, covering the whole island as a normal loess plateau with a total thickness amounting to 35–90 m. The lower layers are typical loess, while those near the surface are sandy loess.

Tunisia was separated from Sicily only by narrow straits. Sardinia and Corsica were connected to each other and almost connected to Italy. Most Aegean islands were continental land (Fig. 57).

Figure 58 summarizes the coastal and shelf environments of the Black Sea around 18 ka BP. At that time, when according to Kaplin and Shcherbakov (1986) sea level was 90 m lower than now and the former shoreline was situated near the shelf break, there was practically no shelf, except in areas over-deepened by tectonic movements, like the southwestern continuation of the Crimea Peninsula and the margin of the modern Danube. As a result, some rivers opened directly into the heads of canyons. At some places, where the continental shelf was very narrow, the heads of canyons and valleys were practically in the coastal wave zone, which promoted highly active slope processes and stimulated landslides. The northwestern part of the shelf was an alluvial plain. The entire watershed was covered by loess. The main Danube channel flowed to the south at that

Figure 57 Land areas exposed due to sea-level lowering in the Mediterranean according to Thiede (1978), modified by Dawson (1992). Winter palaeosea-surface temperatures (°C) have been inferred from planktonic foraminiferal assemblages

73

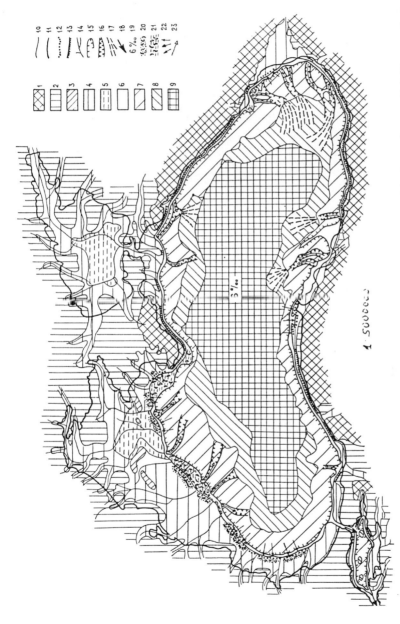

Figure 58 Palaeogeographic map of the Black Sea during the last glaciation (18 ka BP). (1) Mountainous relief, (2) watersheds covered by different types of sediments, (3) loess plateaus, (4) older alluvial and marine coastal plains, (5) alluvial plains formed around 18 ka BP, (6) palaeoshelf, (7) continental slope, (8) continental rise, (9) deep-sea basin, (10) modern shoreline, (11) 18 ka BP shoreline, (12) erosional shores, (13) depositional shores, (14) palaeorivers, (15) shelf margin, (16) submarine canyon, (17) deep-sea fans, (18) drainage from the Black Sea, (19) palaeosalinity, (20) muddy sediments with Dreissena (newexinian, c. 18 ka BP) shells, (21) shell sand with Dreissena shells, (22) sediment movement on continental slope and submarine canyons, (23) sediment drift at near-shore zone (after Kaplin and Shcherbakov 1986)

Figure 59   Possible landscape of an interglacial atoll during a glacial low sea level:
a high limestone cliff (undercut by shorelines corresponding to previous interstadials)
surrounded by a narrow, recent fringing reef (Rurutu Island, French Polynesia)
(Photo 5744, Apr. 1980)

time. The palaeo-Don river crossed the dry bed of the Azov Sea in a
partially oblique direction.

The surface of mid-oceanic islands of volcanic origin was little modified
by the relative sea-level fall, though small ice caps may have formed on
the upper slopes of the highest volcanoes, like the Mauna Kea (Hawaii),
where an area of about 70 km^2 was ice-covered during the last glacial
maximum, with a thickness of as much as 100 m (Porter 1979). More
dramatic topographic changes in elevation occurred to the 400 present-day
coral reef islands listed as atolls by Bryan (1953), which were tranformed
into high limestone islands and attacked by weathering and karstic
processes (Fig. 59).

The Seychelles Islands are a special case. They consist of a bank of coral
reef limestone capping a granite bedrock, which is exposed in high islands
(e.g. Mahé, Praslin). During the last glaciation, the whole Seychelles Bank
was a wide single island, with several valleys of palaeoriver beds reaching
the outer slope or the inner depressions (Kaplin et al. 1986) (Fig. 60).
Weathering and karst processes were active. This situation changed only
about 10 ka ago (Badyukov et al. 1989), when the postglacial sea level
reached –70 to –60 m, penetrating into the inner depressions of the bank
and forming a number of lagoons with mangrove vegetation, which were
subsequently submerged by further sea-level rise.

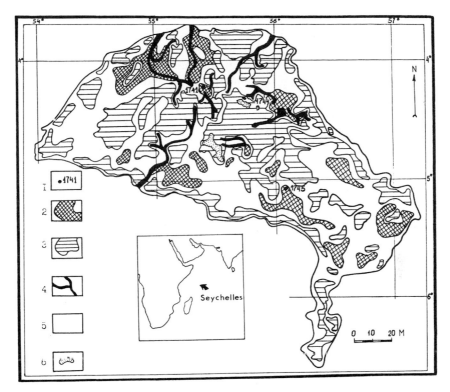

Figure 60 Schematic map of the Seychelles Bank. 1, Core site; 2, palaeolagoon at 50–65 m depth; 3, submerged terrace at 15–45 m depth; 4, palaeoriver valley; 5, sea floor 45–60 m deep; 6, emerged land (from Pirazzoli 1986c)

3.4 LAST GLACIATION CLIMATE AND HYDROLOGY

According to COHMAP Members (1988), at 18 ka BP, when ice sheets were at their maximum, strong anticyclonic air circulation around the Canadian ice sheet was bringing cold conditions to the North Atlantic. Along the southern flank of the ice sheets, strong easterly winds and a sharp temperature gradient were associated with a strengthened jet stream aloft, which extended across North America and east to Eurasia.

North–south temperature gradients were strengthened over Eurasia. This helped to displace the polar front and the westerlies southwards to parts of northwest Africa, the Mediterranean and southern Asia. At the same time, according to modelling by the COHMAP Members (1988), the circum-antarctic sea–ice boundary was about 7° latitude closer to the equator than at present, with a resulting steeper temperature gradient which strengthened southern westerlies, especially across the Indian and the South Pacific oceans.

In tropical areas, lowland terrestrial and sea-surface temperatures have shown changes between glacial and interglacial times much smaller than those at high latitudes: 3–4°C off southeast Africa (Van Campo et al. 1990) and 5°C in Barbados (Guilderson et al. 1994). More detailed reconstruction of the distribution of sea-surface temperature and other boundary conditions for the climate of 18 ka BP have been provided by CLIMAP Project Members (1976).

Rainfall conditions have varied over the continents. In the western United States, the Great Basin was much moister 20 ka ago than it is today. The Middle East seems to have experienced moister conditions similar to those of northern Africa, with a drier climate 18 ka ago (Gasse et al. 1990) and a wetter one by the time that agriculture was developed about 10 ka ago (Webb et al. 1985). Many of these changes were related to changes in the monsoonal climates, which were enhanced by the greater seasonality of the climate between 12 and 5 ka BP. As a consequence, lakes in eastern and northern Africa as well as India, which were dry or shrinking in size 18 ka ago, reached high-water levels between 12 and 5 ka ago.

Spruce and oak forests were absent from Europe because of the cold, dryness and permafrost. The Mediterranean lowlands were treeless. In North America, spruce forest dominated in the Midwest, subalpine parkland in the Pacific Northwest, and tundra in Alaska (COHMAP Members 1988).

The extent of forests in the Atlas Mountains and West Africa was greatly reduced, with dry steppe vegetation in the Mediterranean area. Palaeoclimatic evidence suggests that mean temperatures were 4 to 6°C lower than today in Australia, New Guinea, New Zealand and South America.

The trajectory of major oceanic currents at the time of the glacial maximum was very different from today. In the North Atlantic, for example, winter sea ice extended south to the coast of France. Near the sea-ice border in both the North Atlantic and the North Pacific, sea-surface temperatures are estimated to have been as much as 10°C lower than those at present. The 10°C isotherm of January sea-surface temperature 18 ka ago was generally limited to around the latitude of 40°N, whereas today it can reach 55°N near Ireland (Fig. 61).

The hydrologic balance, which controls global sea level (see Chapter 1), was modified by the huge water masses held back on the continents as ice sheets (some 40×10^6 km^3). Permafrost, which was seasonally ubiquitous at latitudes higher than 30°N in Asia and 40°N in Europe, and permanent almost everywhere above 50°N, often reached several hundred metres in thickness (Fig. 62 (Plate V)).

Figure 61   (*opposite*) January sea-surface temperature at present (a) and 18 ka ago (b), according to Keffer et al. (1988). The heavy solid line is the 10°C isotherm. The crossed line represents the circulation pattern of the present Gulf Stream–North Atlantic Current (a) and the position of the polar front 18 ka ago

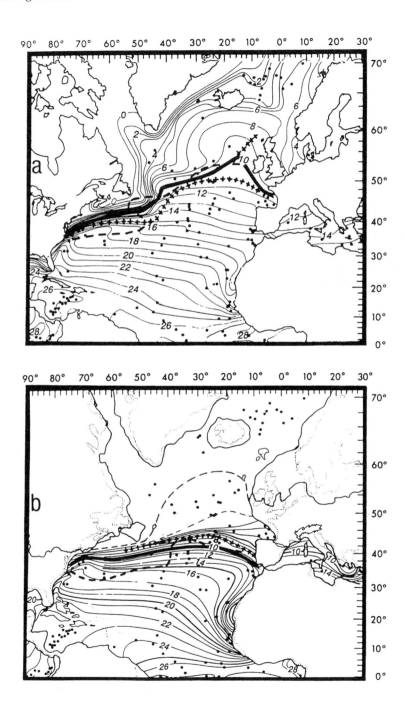

3.5   LAST GLACIATION BIOMASS AND CO_2 EXCHANGES

Carbon, in the form of CO_2, carbonates and organic compounds, is cycled between various reservoirs, atmosphere, oceans, land biota and sediments. The largest natural exchanges occur between the atmosphere, the terrestrial biota and the surface water of the oceans. The marine biota acts as a biological pump, transporting organic carbon as a rain of detritus from the surface waters to deeper layers. Photosynthesis, CO_2 production by plants and microbial respiration contribute to exchanges between terrestrial biota and the atmosphere (Watson et al. 1990).

The most reliable information on past atmospheric CO_2 concentrations is obtained by the analysis of air bubbles from polar ice cores. Measurements on samples corresponding to the last glacial maximum from ice cores from Greenland and Antarctica showed concentrations of 180–200 parts per million (ppm), i.e. 30% less than the pre-industrial (i.e. pre-18th century) value.

A comparison between global terrestrial biomass at the last glacial maximum and at the present time is shown in Table 3. It appears that the total terrestrial biomass carbon during the last glacial maximum was about one-half of the present value, in spite of the fact that the terrestrial area during glacial times was over 10% larger than today, owing to continental shelf exposure at low sea level.

The main causes of CO_2 changes between an interglacial and a glacial period are summarized in Fig. 63. Towards a glacial maximum a huge mass of water is transferred from the ocean (through the atmosphere) towards the ice caps. The cooling ocean stores more carbon, the glacial atmosphere is poor in CO_2 and the terrestrial carbon storage is reduced because of forest destruction and soil erosion. Methane production is low because of reduced wetlands resulting from dryness and cold.

During a deglaciation, on the other hand, meltwater flows back to the ocean. Warming ocean waters release CO_2, which then becomes stored in the expanding forests and soils, and accumulates in the interglacial atmosphere to give a higher CO_2 content. The increasing methane production (resulting from decreasing permafrost and development of wetlands) flows against the net carbon flux (Faure et al. 1993).

Sea-level changes, like methane, have an effect which tends to limit the carbon imbalance of climatic origin: during glacial times, when desert and ice-covered areas increase, new land areas, often productive, are available to vegetation on the continental shelves exposed by the sea-level fall, thus increasing the total terrestrial surfaces by over 10%; during deglaciation, when the global biomass increases, continental shelf areas are submerged by the sea-level rise and the carbon of their vegetation and soils is fossilized in the ocean.

Table 3 Global terrestrial biomass for the last glacial maximum and for the present, according to Faure (1990)

Ecosystem type	Glacial scenario			Present situation (1975)		
	Areas (km² ×10⁶)	Biomass DM[a] (×10¹⁵ g)	Carbon (×10¹⁵ g)	Areas (km² ×10⁶)	Biomass DM[a] (×10¹⁵ g)	Carbon (×10¹⁵ g)
Tropical rain forest	2	84	37.8	10	420	189
Tropical seasonal forest	1	25	11.25	4.5	112.5	50.6
Temperate evergreen forest	1	30	13.5	3	90	40.5
Temperate deciduous forest	1	28	12.6	3	84	37.8
Boreal forest	2	48	21.6	9	216	97.2
Other forest	1	20	9	1.5	30	13.5
Total forest	*8*	*235*	*105.75*	*31*	*952.5*	*428.6*
Woodland and shrubland	4	47.6	21.42	4.5	53.55	24.09
Savanna	36	234	105.3	22.5	146.25	65.8
Grassland	19	30.4	13.68	12.5	20	9
Tundra	12	16.8	7.56	9.5	13.3	6
Desert and semi-desert	35	28	12.6	21	16.8	7.56
Extreme desert (rock,. . .,ice)	42	1.26	0.567	24.5	0.735	0.3
Cultivated land	–	–	–	16	6.4	2.88
Swamp and marsh	4	52	23.4	2	26	11.7
Lake and stream	4	0.08	0.036	2	0.04	0.018
Human area	–	–	–	2	3.2	1.4
Others	1	4	1.8	1.8	7.2	3.24
Total non-forest[b]	*157*	*414*	*186*	*118.3*	*293.5*	*132*
Terrestrial total[b]	***165***	***649***	***292***	***149.3***	***1245.9***	***560.6***

[a] DM = dry matter

[b] The calculations are related to vegetation only. Including soils, the total carbon storage would increase to 968×10¹⁵ g for the glacial scenario and to 2319×10¹⁵ g for the present situation (Adams et al. 1990)

QUATERNARY CHANGES IN THE EARTH SYSTEM

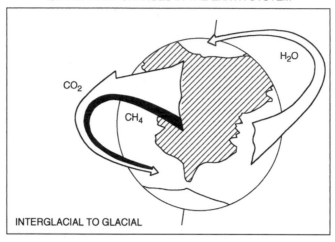

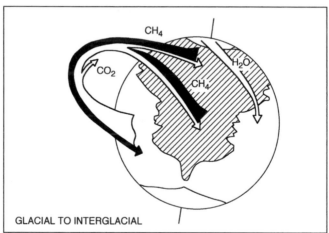

Figure 63   Quaternary changes in the Earth system. Top: interglacial to glacial; below: glacial to interglacial (after Faure et al. 1993)

Chapter Four

Deglacial sea-level changes

4.1 INTRODUCTION

The last deglaciation seems to have started earlier in Antarctica (*c.* 17 ka BP) than in Greenland (*c.* 15 ka BP) (Jouzel et al. 1994; Sowers and Bender 1995). Warming from average glacial conditions to the Holocene was large in central Greenland (*c.* 15°C). This is at least three times the coincident temperature change in the tropics and mid-latitudes. The coldest periods of the last glacial were probably 21°C colder than at present over the Greenland ice sheet (Cuffey et al. 1995).

Increasing evidence supports the idea that the deglaciation occurred in two steps, with a first warming period which peaked at the Bölling (about 13–12 ka BP), and a second warming period after about 10.3 ka BP (i.e. after 11.6 ka ago, calibrated date), separated by a temporary cooling and southward migration of the polar front (Younger Dryas), very marked in North Atlantic areas (Rind et al. 1986) (Fig. 64). However, this cooling was not limited to the North Atlantic and northwest Europe; evidence of it has also been reported from Costa Rica (Islebe et al. 1995), Peru (Thompson et al. 1995), southernmost South America (Heusser and Rabassa 1987), the North Pacific Ocean (Kallel et al. 1988), New Zealand (Lowell et al. 1995), the Gulf of Mexico, the equatorial Atlantic Ocean, the Bengal Fan and the Sulu Sea (references in Kudrass et al. 1991).

The origin of this sudden cooling is not yet fully understood. Broecker et al. (1985, 1989) suggested that it may have been caused by a rapid lowering of salinity in the surface water of the whole northern part of the North Atlantic, following a diversion of the Canadian ice-sheet meltwater flow from the Mississippi to the St Lawrence River. This would have sufficiently lowered the density of the water to prevent it sinking, thus upsetting deep-water circulation in the North Atlantic. Because North Atlantic deep-water sinking is responsible for meridional heat transport to the North Atlantic, a lower rate of deep-water formation would have resulted in cooler conditions during the Younger Dryas. The above interpretation has been discussed by Shackleton (1989) and Edwards et al. (1993). Another possible cause of the

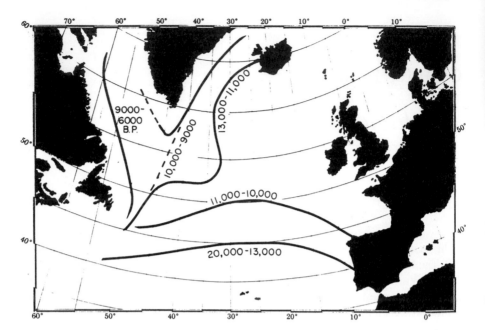

Figure 64   Retreat positions of the North Atlantic polar front from the glacial maximum position 18 ka BP to the modern interglacial location after 6 ka BP (after Ruddiman and McIntyre 1981)

Younger Dryas event may have been the occurrence of massive iceberg discharges (Broecker 1994).

Greenland ice-core records (Johnsen et al. 1992; Grootes et al. 1993) confirm a very rapid transition from the Younger Dryas to the Holocene climate (an increase in temperature of 7°C in 50 years) about 11.6 ka ago. Some evidence suggests an even more rapid transition ($\leq$20 years) (Alley et al. 1993; Mayewski et al. 1993). In Antarctica, however, the temporary cooling is about three times weaker than that observed in Greenland during the Younger Dryas. Also, the last temperature optimum occurred earlier in Antarctica (*c.* 11 ka BP) than in the northern hemisphere (*c.* 6 ka BP) (Jouzel et al. 1994).

The melting histories of the various ice sheets were nonsynchronous and the last deglaciation ended in each place when the former ice sheets had completely melted. This seems to have occurred by 10 ka BP in Scotland (though most of the ice had gone from Britain by 13 ka BP), and around 10 ka BP for the Cordilleran ice sheet, but close to 9 ka BP in the Russian Arctic (Dawson 1992), around 7.5 ka BP in Scandinavia, when the last parts of northern Sweden were uncovered near the centre of the former ice sheet, and around 6 ka BP for the Laurentide ice sheet. In Baffin Island, where the melting of an ice cap is estimated to have contributed to about

0.5 m of global sea-level rise in the last 6–7 ka, permanent glaciers still exist today in the Barnes area (Bloom 1971). In Antarctica and Greenland, only the outer parts of the ice domes melted, owing to the rise in sea level and to the new equilibrium profiles to which both the ice caps had to adapt after the displacement of the new boundaries between continental ice and sea water.

Deglacial history can be studied in formerly glaciated areas, but the ice masses involved can be estimated only with great uncertainty. Relative sea-level changes can be inferred only for postglacial times, because marks of former shorelines on ice vanished with ice flowing and melting. This is why accurate determinations of relative sea-level positions during glacial times, generally so rare, are virtually impossible in formerly glaciated areas. In such areas, local sea-level histories can be reconstructed only after deglaciation, from features left at the marine limit or by terrestrial organisms on land uncovered by ice, or from deposits like sandur deltas, which are related to the position of the water level beyond the retreating ice margin at the time they were active. It is not surprising, therefore, that most relative sea-level curves reported from formerly glaciated areas rarely go beyond the Holocene period and, when this is the case, only for a few thousand years in the most favourable cases. Near the centre of ancient ice sheets relative sea-level histories are even shorter. In other words, there is no way of accurately measuring sea-level changes and vertical isostatic movements prior to deglaciation; in these areas estimates can only be based on assumptions and modelling, or on proxy data, rather than on direct observation.

4.2 MODELLING RESULTS

Global isostatic models may predict relative sea-level changes forced by deglaciation. The models are usually based on the mathematical analysis of the deformation of a viscoelastic Earth produced by surface mass loads. Global isostatic models, which brought a conceptual revolution in sea-level research, were initiated in the early 1970s by the pioneering work of Walcott (1972), Peltier (1974, 1976) and Cathles (1975). They were first used by Peltier and Andrews (1976), on the assumption that meltwater from ice sheets was distributed uniformly over the global ocean (Peltier 1990).

In predicting sea-level variations, a melting history has to be assumed for all the continental ice loads that existed at the time of the last glacial maximum. Some melting histories assumed by global isostatic models (which may differ notably) have been summarized by Pirazzoli (1991, p. 27).

Isostatic effects of deglaciation have also been modelled on a regional scale, as in Atlantic Canada (Quinlan and Beaumont 1981, 1982), Fennoscandia (Fjeldskaar 1991), oceanic islands (Nakada 1986), the Australian region (Nakada and Lambeck 1989), Japan (Nakada et al. 1991), the British Isles (Lambeck 1993a,b), Greece and southwestern Turkey (Lambeck 1995), the Adriatic Sea (Lambeck and Johnston 1995) and the Atlantic and Channel coasts of France (Lambeck 1996).

In a clear, didactic paper, Lambeck (1993c) has distinguished the *near-field* locations (defined as those that occur within the limits of the former ice sheets) from the *ice-margin* sites (for locations near the former ice margins), the *intermediate-field* sites, and lastly the *far-field* sites (defined as those well away from the influence of the former ice sheets, such as locations along the Australian margin or Pacific Ocean islands). What is remarkable in most global isostatic models is that the vertical land movements caused by deglaciation are not limited to formerly glaciated areas and to nearby regions, as proposed by Daly (1934) (Fig. 4), but are assumed to extend more or less all around the globe.

In the near-field sites, which correspond to Zone I of Clark et al. (1978) (see Fig. 6), the dominant contribution to sea-level change comes from ice-load effects, and the characteristic late-glacial and postglacial relative sea-level curve is one of almost exponential fall up to the present because of rising land.

At ice-margin sites (Transition Zone I–II in Fig. 6), the relative sea-level change is characterized by an initially rapid fall during the late-glacial stage, followed by a period of relative stability, then by a rise in level, and finally by a more or less uniform sea-level fall to the present position.

The intermediate-field sites (Zone II in Fig. 6) correspond to the peripheral bulge around former ice margin which tends to subside in late-glacial and postglacial times, to compensate the uplift in nearby formerly glaciated areas. In these sites the relative sea level continues to rise even when deglaciation has ceased, though at gradually decreasing rates.

In far-field sites, lastly, glacio-eustatic changes in sea level are considerably greater than glacio-isostatic and hydro-isostatic effects, and relative sea-level rise only predominates during the deglaciation period, often followed by a slight relative sea-level fall of hydro-isostatic origin during the late Holocene.

4.3 REGIONAL CASE STUDIES

The number of relative sea-level curves reaching late-glacial or full-glacial times is quite limited in the literature. Some published curves have been assembled below, the main criterion for selection being that they go back, if not always to the last glacial maximum, at least to 13 000 or 14 000 radiocarbon years BP. These curves have been reduced at the same scale and reproduced in Figs. 65–69. Unfortunately, these curves represent only a few coastal areas of the world. They are therefore insufficient to give a detailed overview of sea-level changes for the last 20 ka, such as the review carried out by Pirazzoli (1991) for the Holocene period. However, a comparison between the various curves will provide at least a qualitative idea of the type of results which can be reached with certain approaches. In addition, unexplained discrepancies may indicate the resolution which can be obtained in this kind of research and the improvements which are needed, especially when uncertainty margins have not been indicated by the authors.

4.3.1 Near-field and ice-margin sites from North America

The curves in Fig. 65 are from the Canadian Beaufort shelf (curve A), eastern Canada (curves B–D) and part of the east coast of the USA. They are all of the ice-margin type, except B, which is a typical near-field curve.

A dozen radiocarbon-dated peat and peaty clay samples from geo-technical boreholes in the Canadian Beaufort continental shelf, off Tuktoyaktuk Peninsula, have been used by Hill et al. (1985) to reconstruct a relative sea-level curve covering the last 30 ka (Curve A; only the last 22 ka are shown in Fig. 65). Seismic profiles were used to determine the environment of deposition. Curve A shows a history of 140 m of sea-level rise since 27 ka BP, 35 m of which can be related to subsidence (basin subsidence, sediment loading and consolidation subsidence), the remaining 105 m being the consequence of glacio-eustatic effects. A relative sea-level fluctuation with a minimum to −70 m between 20 and 10 ka BP has been interpreted as resulting from the displacement and collapse of the peripheral bulge caused by the ice load. This implies a location on the crest or on the distal side of the peripheral bulge, facing away from the ice load, and an ice readvance of late Wisconsin age.

From the opposite side of the Canadian ice shelf, curve B was obtained by Locat (1977), using 11 radiocarbon-dated shell samples, in the Baie des Sables–Trois Pistoles area, where the marine limit (about 13.5 ka BP) is found between +120 and +166 m. Curve (band) C by Brookes et al. (1985) summarizes the relative sea-level history in southwestern Newfoundland, where the marine limit is reported at +44 m around 13.6 ka BP, and sea level remained below the present level since slightly less than 10 ka BP. In both southwestern and southeastern New Brunswick, emergence com-menced approximately 16 ka BP according to Scott and Medioli (1980), who used marine–fresh water transitions in sediments on raised basins formerly below sea level as sea-level indicators. After a most rapid phase between 16 and 15 ka BP, the rate of emergence appears to have lessened to 5 mm/yr (curve D). In coastal Maine, according to Kelley et al. (1988), an initial transgression, which closely followed retreat of the ice front, was accompanied by deposition of glacio-marine sediments (largely mud). Following this deposition, the relative sea level (curve E) is believed by

Curve	Location	Reference
A	Canadian Beaufort shelf	Hill *et al.,* 1985
B	St. Lawrence River, Québec	Locat, 1977
C	West Newfoundland	Brookes *et al.,* 1985
D	New Brunswick	Scott & Medioli, 1980
E	Maine	Kelley *et al.,* 1988
F	Maine	Newman *et al.,* 1971,1980
G	Boston area	Newman et *al.,* 1971,1980
H	Virginia	Newman *et al.,* 1971,1980
I	Iona Island	Newman *et al.,* 1971,1980
J	New York City	Newman *et al.,* 1971,1980

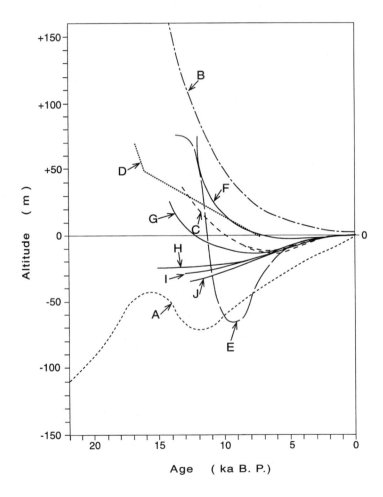

Figure 65   Deglacial relative sea-level changes in North America: I

Kelley et al. (1988) to have fallen rapidly to an erosional unconformity developed on the former sea floor, where a pronounced (undated) shoreline was found at 60 m depth; the sea-level rise following this lowstand position is believed to have slowed between 4.0 and 2.5 ka BP.

Relative sea-level curves F to J in Fig. 65 were proposed by Newman et al. (1971) in a pioneer paper discussing glacial peripheral bulge development in North America and northern Europe. Some of these curves are based on only a few points, however, and should be considered as indicative; the main conclusion by Newman et al. (1971) was that curves F to J diverge going back in time, much of the delevelling being caused by postglacial isostatic rebound in Maine and eastern Massachusetts.

4.3.2 Intermediate-field sites from North America

The curves in Fig. 66 show data available from the Atlantic continental shelf of the USA and off Barbados. Most of the coasts of the USA and the West Indies (as well as those in western Europe) can be considered as intermediate-field areas. Curve A by Curray (1965) shows approximate means of dates compiled from the continental shelves of the USA (Atlantic, Gulf and Pacific sides). The maximum late-Wisconsin lowstand of sea level is not well-established, but it was probably at least 110 m (−124 m is Curray's estimate), and there are very few reliable dates older than 10 ka. Curve B was obtained by Milliman and Emery (1968) from some 80 radiocarbon dates of samples collected from the Atlantic continental shelf of the USA and considered to be eustatic, with a lowstand at −130 m about 15 ka BP. Dillon and Oldale (1978) found evidence of differential subsidence in part of the continental shelf south of Long Island and obtained curve C, between 20 and 8 ka, as representative of eustatic sea level for the "US East coast". Curves A and B differ considerably between 17 and 9.5 ka BP from curve D, by Blackwelder et al. (1979), which is based only on in-place lagoonal and salt-marsh material cored from the continental shelf between Delaware and northwest panhandle Florida.

Curves A, B, C and D in Fig. 66 have been obtained following a similar approach, using shelf data from very wide areas. Such an approach is inaccurate, because it does not take into account glacio-isostatic and tectonic effects which may change with the area investigated, and hydro-isostatic effects which vary along each shelf transect perpendicular to the coast, depending on the distance from the present-day shoreline and water depth.

Much more acceptable is the approach followed by Fairbanks (1989) for curve E and by Bard et al. (1990b) for curve F, using the same samples. Fairbanks (1989) was the first to obtain a continuous and detailed record of sea-level change from a geographically limited area during much of the last deglaciation period. Sixteen cores were drilled off the south coast of

Curve	Location	Reference
A	U.S. continental shelves	Curray, 1965
B	U.S. Atlantic shelf	Milliman & Emery, 1968
C	Eastern U.S. shelf	Dillon & Oldale, 1978
D	Southeast U.S. Atlantic shelf	Blackwelder *et al.*, 1979
E	Barbados	Fairbanks, 1989
F	Barbados	Bard *et al.*, 1990 b

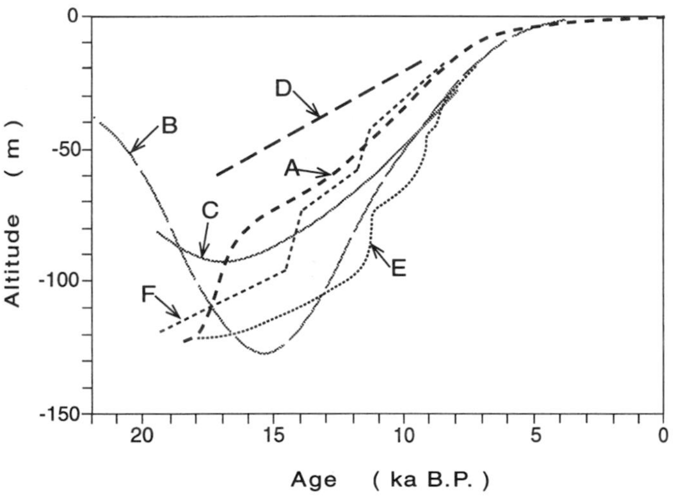

Figure 66   Deglacial relative sea-level changes in North America: II. The ages of curve F have been calibrated, whereas all other curves are in uncalibrated radiocarbon dates

Barbados, through an almost continuous sequence of the reef-crest coral *Acropora palmata* between 17.1 and 7.8 ka BP. *A. palmata* is generally restricted to the upper 5 m of water (below the spring low-tide level). After correction for an uplift rate of 0.34 mm/yr for the south coast of Barbados (this rate was deduced from the present-day elevation of Pleistocene raised reef terraces), curve E was constructed. It shows a rise in sea level of 121 ± 5 m since 18 ka BP, with two periods of more rapid rise around 12 and 10 ka BP.

Bard et al. (1990b) repeated the dating of all Barbados samples by accelerator mass spectrometry (AMS) of ^{14}C, as well as by thermal ionization mass spectrometry (TIMS) of U/Th, thereby obtaining, for the first time, a calibration of radiocarbon ages older than 13 ka BP. This calibration showed that ^{14}C ages are systematically younger than U/Th ages, which are more accurate because they accord with the dendrochronological calibration (see Section 2.4). In Fig. 66, the ages of curve E are therefore calibrated (i.e. expressed as ka ago), while the ages of the other curves in

Figs. 65–69 are in radiocarbon ka BP. The occurrence of two large sea-level jumps can be shown more accurately from curve F thanks to the greater precision of the U/Th chronology: after a first period of sea-level rise beginning at least 19 ka ago at an average rate of 5 mm/yr, the first jump corresponds to a dramatic acceleration to about 37 mm/yr, which occurred about 13.5 ka ago, followed by a decrease to about 8 mm/yr; the second abrupt change was an acceleration from 8 to 25 mm/yr, around 11 ka ago (Bard et al. 1990a), which may have lasted as long as 1.7 ka (Bard et al. 1990b). The rates of the corresponding two meltwater pulses were higher than 10^6 km^3 of continental ice per 100 years.

Can sea-level curves E and F (which are the same, differing only for the time-scale calibration) be considered globally representative, as claimed by Fairbanks (1989)? In fact, where a correction for tectonic uplift has been applied, it does not include glacio-isostatic and hydro-isostatic effects. Curves E and F are therefore representative of the relative sea-level history in an intermediate-field site (Zone II in Fig. 6). According to global isostatic models, they cannot represent the global situation more than any other local sea-level curve obtained from field data, in the absence of an adequate isostatic correction, however accurate and reliable the field data may be. On the other hand, the various acceleration and deceleration periods shown by curve F can hardly result from tectonic movements and are therefore more likely to be of global significance.

4.3.3   Ice-margin and intermediate-field sites from Europe

In Fig. 67, some relative sea-level curves from western Europe have been summarized. Curve A from Spitsbergen is of the ice-margin type, whereas curves B to F are typical of intermediate-field areas.

The Svalbard Archipelago is located at the northwestern margin of the shallow Barents shelf, where the occurrence of a wide ice sheet at the time of the glacial maximum has been postulated (e.g. Solheim et al. 1996). Svalbard was characterized by a marked asymmetry of the late-Weichselian ice cover, which was large in the east but much more moderate on the west coast (André 1993). Accordingly, glacio-isostatic rebound (shown by several emergence curves, some of which have been compared by Héquette, 1988) has uplifted Holocene beaches in the east to over 100 m (Salvigsen 1981), whereas on the west coast the uplift may be less than 20 m (Pirazzoli 1991, plate 1). A relative sea-level curve from 13 ka BP (A) has been proposed by Forman et al. (1987) for the northwestern coast of Spitsbergen; it shows a fall in the relative sea level of at most 5 mm/yr during the initial period of ice unloading (from prior to 13 ka BP until 10 ka BP), interrupted by at least three short periods of stillstand or possible transgression. The local sea level fell rapidly (30 mm/yr) after 10 ka BP, coinciding with the final deglaciation of Spitsbergen, and intersected modern sea level by 9 ka BP. Later, a mid-

Curve	Location	Reference
A	Spitsbergen	Forman *et al.*, 1987
B	NW Europe	Mörner,1980
C	W of Belle Ile (France)	Pinot, 1968
D	Roussillon (France)	Labeyrie *et al.*, 1976
E	SW France	Aloïsi *et al.*, 1978
F	French Riviera	Dubar & Anthony, 1995

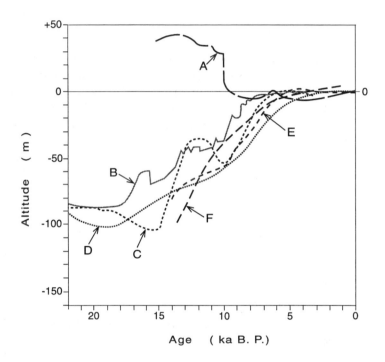

Figure 67   Deglacial relative sea-level changes in Europe

Holocene transgression brought sea level up to, or above, its present level around 6 ka BP.

Curve B by Mörner (1980) was deduced from about 40 shoreline sectors from the Kattegatt region (west coast of Sweden), a marginal area of the Fennoscandian uplift, and checked and confirmed against other north-western European records (Shennan 1989). Considered by Mörner (1980) as "eustatic", or as "regional eustatic" (Mörner 1979–1980), curve B has often been used as a reference, especially in northwestern Europe.

Curve C by Pinot (1968) was interpreted from morphological evidence observed on the continental shelf west of Belle Ile (France), but local radiometric control was missing. On the continental shelf of Roussillon (southwestern France), several mollusc shell samples, collected from borings, have enabled Labeyrie et al. (1976) to construct a relative sea-level curve covering the last 30 ka. This curve (D), which is typical of intermediate-field sites, shows a minimum stand at about −100 m at the time of the last glacial maximum, followed by a gradual rise to the present level, with a deceleration during the last few thousand years. A similar shape is shown by curve E (Aloïsi et al. 1978), which includes data from many borings in the coastal plain and beach sites of Roussillon, as well as palynological observations from lagoonal sediments. Curve E culminates at +2 m with beach deposits dated between 5 and 4 ka BP near the base of a raised notch (which, however, may be older than the beach deposits) near Cap Romarin, and shows almost no change (maintaining present sea level) during the last 2 ka. Lastly, curve F, proposed for the French Riviera by Dubar and Anthony (1995), shows a rapid sea-level rise up to around 5 ka BP, with little change thereafter. A late Holocene relative sea-level stand slightly higher than the present one is interpreted by Dubar and Anthony (1995) from river-mouth deposits; however, this is in contradiction to the biological evidence of continuous sea-level rise between 4.5 and 1.5 ka BP provided by Laborel et al. (1994) on the rocky coasts of southwestern France, and to the complete lack of evidence of emergence reported by Morhange (1994) from rocky cliffs in the same area. Near Cassis (southern France), the access to the Cosquer cave at −37 m was closed by the sea-level rise about 9 ka BP (Sartoretto et al. 1995). Prehistoric paintings on the walls of the cave are intact above the present sea level, but completely destroyed below, demonstrating that the present sea level is the highest reached during the Holocene in that area (Collina-Girard 1995).

4.3.4 Intermediate-field sites from the Black Sea and the Caspian Sea, far-field sites from West Africa

Relative sea-level curves from the Black Sea, the Caspian Sea and West Africa have been assembled in Fig. 68. Water exchanges between the Mediterranean, the Black Sea and the Caspian Sea were mainly controlled by the variations in their surface levels, due to eustatic, climatic and tectonic changes, and the elevation of the Bosporus sill (at about −35 m, whereas the present-day level of the Caspian Sea is about −28 m). In late glacial times the Black Sea and the Caspian Sea were still isolated from the Mediterranean, and the Bosporus (or possibly another strait crossing the Sapanca–Sakarya area, as suggested by M. Ozdogan, personal communication) was discharging brackish Black Sea water into a lowered Mediterranean.

The hydrologic balance of the Caspian Sea differed from that of today, as suggested by presently inactive river beds in central Asia, such as the Ouzboï, which once flowed into the Caspian. At certain times, when the level of the Caspian was higher than that of the Black Sea, a discharge

Curve	Location	Reference
A	Black Sea	Serebryanny, 1982
B	Caspian Sea	Chepalyga, 1984
C	Sénégal	Faure & Elouard, 1967
D	Ivory Coast	Martin & Delibrias, 1972
E	Off southern Gabon and Congo	Giresse *et al.*, 1986

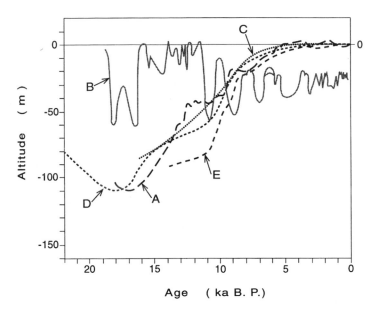

Figure 68   Deglacial relative sea-level changes in the Black Sea, the Caspian Sea
and West Africa

system was activated in the Manych–Kuma area, decreasing the differences
in level between the two basins. Sea-level fluctuations in the Black Sea may
therefore have been different from those in the Mediterranean or in the
Caspian Sea, but were in some way linked to each other beyond certain
altimetric limits.

Curve A (Fig. 68), proposed by Serebryanny (1982) for the Black Sea,
shows a gradual sea-level rise from –110 m (17 ka BP) to the present sea
level (about 5 ka BP), with two major interruptions, at about –45 m
(between 12 and 10 ka BP) and –20 m (around 9–8 ka BP). Several minor
sea-level fluctuations which appear in curve A may be due to the fact that
the data compiled by Serebryanny (1982) come from different sites around
the Black Sea, which may have had different tectonic histories.

In the Caspian Sea, which was at a higher level than the Black Sea
during early deglacial times, many sea-level fluctuations of great amplitude
have been proposed by Chepalyga (1984) (curve B). Detailed information

on the evidence for these oscillations is not available, but during historical times, according to Julian et al. (1987), the level of the Caspian Sea was high around 1.7 ka BP, low (–30.5 m) at 0.7 ka BP and high again (–25.3 m) in the 18th century. The last peak of high sea level (AD 1929) was followed by a rapid 3 m fall, stabilization in 1975–1977, and a new 1.5 m rise during the following decade (Ignatov et al. 1993).

The evidence available from West Africa usually provides consistent sea-level histories. Curve C is an early sea-level curve, proposed as eustatic by Faure and Elouard (1967), following eustatic models propounded at the time by Shepard and Jelgersma (in regions slowly subsiding), but under the assumption that the west coast of Africa was rising at a rate of 0.5 mm/yr. Very similar to curve C, but more complete and based on local data (at least for the end of the Pleistocene), is the relative sea-level curve D, produced by Martin and Delibrias (1972) for the Ivory Coast. Curve D is based on radiocarbon dates of samples collected by coring or dredging at depths between 43 and 100 m on the continental shelf. It shows a lowstand estimated at about –110 to –115 m for the last glacial maximum and a gradual postglacial sea-level rise, with a marked acceleration around 11–10 ka BP. Finally, relative sea-level curve E was obtained by Giresse et al. (1986) from the southern Gabon and Congo shelves, mainly using radiocarbon-dated samples of cored mangrove peats. The delay of 1–2 ka of curve E compared with curve D was ascribed by Giresse et al. (1986) to an oscillation of the oceanic geoid surface.

4.3.5 Far-field sites from East Asia

Some relative sea-level curves from east Asia are assembled in Fig. 69. Curve A, proposed by Fujii and Fuji (1967) for "the Japanese Islands", is based on over 40 radiocarbon dates and subsurface geological data from several coastal areas of Japan; it shows a rapid rise from –130 m about 18 ka BP to a few metres above the present sea level around 6 ka BP, with a stagnation of sea level around 12–10 ka BP at a depth of about –40 m. Sea-level band B–B', was proposed for "Japan" by Fujii and Mogi (1970). A tentative fluctuation reaching the depth of a group of undated submarine terraces at –10 to –20 m has been interpreted around 14 ka BP by Fujii and Mogi (1970), but the occurrence of such a sea-level fluctuation has not been confirmed by subsequent work. Curves C and C' delimit an envelope of values from the Atlantic shelf of the USA containing curve B proposed by Milliman and Emery (1968) in Fig. 66; according to Emery et al. (1971), the envelope C–C' also contains most of the depths and radiocarbon dates of shallow-water relict shells dredged from the floor of the East China Sea and adjacent area; this is, however, insufficient to provide precise local relative sea-level histories in the East China Sea.

Curve	Location	Reference
A	Japan	Fujii & Fuji, 1967
B-B'	Japan	Fujii & Mogi, 1970
C-C'	East China Sea	Emery et al., 1971
D	East China Sea	Yang & Xie, 1984
E	Eastern China	Zhao & Zhao, 1986

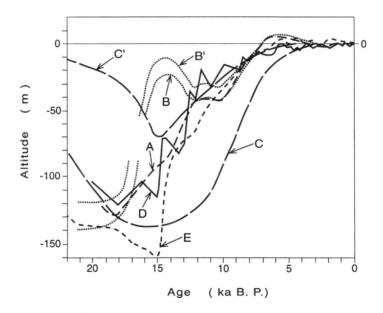

Figure 69   Deglacial relative sea-level changes in east Asia

Relative sea-level curve D was proposed by Yang and Xie (1984) for the "shelf of the East China Sea", using 26 radiocarbon-dated samples with littoral, shallow-sea and terrestrial facies; the serrate aspect of curve D is due to the fact that the various reconstructed sea-level positions, devoid of uncertainty margins, have been connected by straight segments.

Lastly, relative sea-level curve E was constructed for "Eastern China" by Zhao et al. (1982), using over 60 radiocarbon dates of samples collected from wide coastal and shelf areas of the Bohai Sea, the Huanghai Sea, the East China Sea and even northern Taiwan (see also Zhao and Zhao 1986).

In conclusion, most of the sea-level curves in Fig. 69 have been constructed, like curves A–D for the continental shelves of the USA in Fig. 66, with composite data coming from very wide coastal and shelf areas, which have probably been affected by significant differential vertical displacements since the last glacial maximum. They can therefore be considered only as broadly indicative.

4.4 A GRADUALLY RISING OR A FLUCTUATING SEA LEVEL?

Major climatic changes (e.g. glacial–interglacial cycles) which lead to substantial variations in the world water balance, such as those which occurred between about 18 and 6 ka BP, necessarily imply major worldwide sea-level changes. A long-debated point is whether minor sea-level fluctuations may also have occurred during the last glacio-eustatic sea-level rise, because minor global climatic oscillations, or localized climatic changes, may be expected to produce more or less significant sea-level fluctuations.

Geomorphological studies have shown that many of the world's beaches, particularly on oceanic coasts, have received sand or shingle washed in from the sea floor during and since the last major marine transgression. Widespread shoreward drifting then took place as the sea advanced across nearshore shoaly topography, and waves washed sediment on to the shore, forming beaches and dunes. Sandy barriers such as those backing the Ninety Mile Beach in southeastern Australia show evidence of substantial progradation in Holocene times as the result of shoreward movement of the sea floor sand; Chesil Beach in southern England is a shingle barrier that formed and was driven shoreward as the sea rose, with a lagoon on its landward side backed by a submerging, indented hinterland coastline that has never been exposed to the open sea (E.C.F. Bird, personal communication, 1996).

Such shoreward movement of sediment may have been facilitated by minor oscillations of sea level (stillstands or brief episodes of sea-level fall) as the deglacial marine transgression proceeded, because this would have been more effective than a single movement during a gradual sea-level rise. The statistical procedure of fitting curves of sea-level rise to a scatter of dated depth samples, without taking into account adequate uncertainty ranges, inevitably produces a smoothly rising curve, and could actually obliterate evidence of minor oscillations during the marine transgression which took place since the last glacial maximum (Bird 1996).

The possibilities of identification of past sea-level fluctuations depend on the duration of the event, the rate of long-term sea-level change predominant before and after the oscillation, the time scale investigated, and especially the resolution of the sea-level indicators and dating methods used. The problem may be faced at three different scales: global, regional and local.

The global approach does not imply, as some authors apparently still believe, that if a eustatic oscillation occurred in the past, it must have produced the same sea-level fluctuation everywhere around the world; nor does evidence of the absence of a sea-level fluctuation at a site necessarily mean that such fluctuation could not have taken place at other sites. Such a rigid global approach would probably be misleading, owing to the various isostatic, tidal, climatic and other factors already mentioned; if a sea-level fluctuation of global origin occurred, its apparent amplitude may not have been the same everywhere, and it could also be masked by other phenomena at certain sites.

Over a regional scale, similar (though not identical) relative sea-level oscillations may have occurred during the same time interval throughout wide areas; isostatic responses to changes in ice or water load and the corresponding geoidal responses, together with their interactions with glacio-eustatic changes, are indeed to be included among the possible causes of such regional oscillations, especially near former ice margins. Other sources of regional relative sea-level oscillations may be due to (seismo)tectonic processes, particularly near active plate boundaries (see Section 5.3). Changes in the regional ocean-surface topography (owing to changes in ocean circulation or in climate) may also produce small sea-level fluctuations over wide areas, especially when they imply displacements of major oceanic currents, changes in trade winds, or interruption/activation of upwelling processes.

Over a local scale, possible causes of relative sea-level oscillations are even more numerous, because they include, in addition to all global and regional causes, many kinds of possible local vertical displacements, resulting from small-scale hydro-isostatic, tectonic or volcanic processes, sediment compaction, tidal changes, and climatic changes.

Fluctuations in the relative sea level are therefore likely to have occurred in many areas during and after the deglaciation period. However, extrapolation of these fluctuations to regions other than those where they have been documented is generally to be avoided. The Younger Dryas event may provide a good example of the different signatures which may be left by significant eustatic sea-level fluctuations in areas affected by divergent isostatic movements.

4.5   THE YOUNGER DRYAS SEA-LEVEL CHANGE

The Younger Dryas climatic change is probably one of the most significant potential producers of a global sea-level fluctuation which has occurred since the last glacial maximum. As this climatic event caused not only an interruption in ice-sheet melting processes, but also a glacial readvance, it can be expected to have caused a major global sea-level oscillation. A rapid check of Figs. 65–69 does not distinguish any general fluctuation around 11 ka BP. Nevertheless, a closer inspection of the various sea-level curves, taking into account the research methods applied by their authors, may provide some interesting indications.

Among the curves in Figs. 65–69, the most precise sea-level information is probably that provided by curves E and F in Fig. 66 with data from

Barbados, which used different dating methods for the same sea-level samples. Several U/Th dates of *Acropora palmata* in the interval between 12 and 10 ka BP enabled Bard et al. (1990a,b) to recognize an abrupt break in sea-level rise (calculated from slopes of the linear regressions) about 11 ka ago. This break took the form of an acceleration from 8 to 25 mm/yr around 11.5 ka ago, followed by a decrease to 9 mm/yr around 10.5 ka ago, and is equivalent to a sea-level oscillation of total amplitude of about 17 m. Such an oscillation, simply consisting of acceleration and deceleration during a prolonged period of rapid sea-level rise, could hardly be detected with less accurate sea-level indicators, such as samples that may have been displaced or collected with poor vertical resolution from continental shelves.

An acceleration in sea-level rise slightly before 10 ka BP has also been reported by other authors, though with less accuracy, where enough reliable samples from the period considered were available, for example in SW France (Fig. 67, curve E), the Black Sea (Fig. 68, curve A), the Ivory Coast (Fig. 68, curve D), the Congo and Gabon shelf areas (Fig. 68, curve E). On the ice-margin area of West Spitsbergen, the superimposed effects of the glacial and eustatic components of the Younger Dryas event seem to have caused a rapid relative sea-level fall of about 30 m around 10 ka BP (Fig. 67, curve A). The period immediately following the Younger Dryas (*c.* 10 ka BP) may have been characterized by the opposite effect: rapid sea level fall in West Spitsbergen (Fig. 67, curve A), due to accelerated glacial unloading following the Younger Dryas readvance.

The best sites to measure the amplitude of the sea-level oscillation related to the Younger Dryas event are at ice-sheet margins where the glacio-isostatic uplift rate was of the same order as the eustatic rise rate around 11 ka BP. This is the case, for example, on certain sectors of the Atlantic coast of Norway. The amplitude of the oscillation will change from place to place, owing to a complex interplay between glacial isostasy and eustasy, where the magnitude of each component changes throughout the time interval considered. In northern Norway, according to Marthinussen (1961), there is evidence of the occurrence of a sea-level fluctuation of at least 11 m in amplitude between 11 and 10 ka BP; in central Norway, near Bjugn, a marked increase in the rate of relative sea-level fall, from 2 mm/yr before 11.5 ka BP to 3 mm/yr until the end of the Younger Dryas, was observed by Kjemperud (1986); in western Norway, shore-level displacements, presumably resulting from the Younger Dryas glacial advance, have caused a relative sea-level oscillation of at least 11 m at Jæren, 13 m at Yrkje and 12 m at Sotra (Anundsen 1985); and in southwest Norway there was a 7–10 m relative sea-level rise in late Alleröd and early Younger Dryas times, followed by a rapid relative sea-level fall of 16 m throughout Younger Dryas and early Pre-Boreal time (Thomsen 1982; Bird and Klemsdal 1986).

4.6   IMPACTS OF PAST SEA-LEVEL
RISE ON COASTAL SYSTEMS

Any relative sea-level rise not only displaces the shoreline kinematically to a new level, but also modifies the zones exposed to wave action and consequently longshore sediment transport and deposition. Landward migration of a shoreline results in land loss by submergence and salinity increase. Sea-level rise also tends to submerge valleys and cause the deposition of sediment on coastal plains, where the reduced river slope favours the development of peaty and marshy areas. Beaches are eroded and submerged.

During a slow sea-level rise, barrier islands may move landwards, but fringing coral reefs may become barrier reefs offshore.

During a rapid sea-level rise, on the other hand, even coral reef growth comes to an end, and all coastal features are submerged. The sudden rapid sea-level rise evidenced between 14.2 and 13.8 ka ago in Barbados (see Fig. 66, curve F) was large and rapid enough to create a hiatus on the barrier reef of Tahiti (Bard et al. 1996) and to preserve shallow marine deposits from reworking off the east coast of Taiwan, in spite of a locally very rapid rate of tectonic uplift, of the same order as the average deglacial eustatic rise (about 8 mm/yr) (see Fig. 118).

At intermediate-field and far-field sites, even with less rapid rates of sea-level rise, many fossil coastal systems have remained submerged and have been preserved underwater on continental shelves. Fluvial systems extending across exposed continental shelves during periods of lowered sea level and buried by the sea-level rise have been recognized on the continental shelf of Louisiana (Suter 1986) and off several other modern deltas and river mouths. Evidence of fossil lagoons submerged by the sea-level rise between 8.5 and 8.0 ka BP have been reported from the continental shelf south of Dakar, Sénégal (Dumon et al. 1977) and from the Adriatic Sea off Ravenna (Colantoni et al. 1990). Much work remains to be done on the evolution of formerly emerged continental shelf areas during the last glaciation and on the processes accompanying the subsequent sea-level rise.

4.7   PALAEOMONSOONS

According to palaeoclimatic data and general circulation models, the orbitally induced increase in summer-time solar radiation 12 ka to 5 ka ago increased the thermal contrast between land and sea, thus producing stronger summer monsoons, which raised ground-water and lake levels in regions that are arid today (COHMAP Members 1988). What is recorded, at least in marine sediments of the Arabian Sea, is in fact a non-linear response of the monsoonal climate to the variations in intensity of solar insolation during summer, with a series of several distinct, rapid events, each of less than 300 years' duration (Sirocko et al. 1993). This had the effect of retaining on the continents a fraction of the ice-sheet meltwater, possibly slightly decelerating the glacio-eustatic sea-level rise. Subsequently, this water was gradually transferred to the ocean during the arid period of the last 5 ka. The sea-level equivalent of water from such lakes and moisture in the soil may have been almost negligible (see Table 1); but the impact on sea level of monsoon rainfall held back as ground water may have been significant, though it cannot be quantified. According to Cailleux (1969, p. 490), for example, in the Sahara region the phreatic level would have dropped by between 10 m and 100 m, depending on the area considered, during the last 6 ka. Detailed information of this kind is missing from most other areas of potential interest.

Chapter Five

Relative sea-level changes in the late Holocene

5.1 DELTA AND CORAL REEF DEVELOPMENT

Rates of relative sea-level change decelerated considerably during the period from 10 to 6 ka BP. This took place in glacio-isostatically uplifting regions, as well as in intermediate-field and far-field areas (Pirazzoli 1991) and had important geomorphological implications, especially for the development of deltas and coral reefs.

The development of small glacio-marine deltas situated near the margin of retreating ice sheets was not an unknown phenomenon during late-glacial times (Andrews 1986) (Fig. 70). However, most Holocene deltaic sequences began to accumulate systematically only when the rate of fluvial sediment input overtook the declining rate of sea-level change along coasts. As shown by Stanley and Warne (1994), who analysed data from 36 world deltas for which clearly identifiable dated basal or near-basal sections were available, this process occurred on a worldwide basis, on tropical deltas as well as on temperate or high-latitude ones. The oldest radiocarbon age available at or near the base of the deltas considered by Stanley and Warne (1994) ranged from about 8.5 to 5.5 ka BP, with a modal age of 7.0 to 7.5 ka BP. Outside uplifting areas, basal delta deposits tend to be younger in a landward direction, primarily in response to the rise in the relative sea level.

Modern reef morphology depends mostly on the depth and shape of the pre-Holocene foundations and on the local postglacial relative sea-level history. No single reef is known to have kept pace with sea level through-out the entire rise since the last glacial maximum. Sooner or later, all Late Pleistocene reefs gave up during one of the acceleration phases of the postglacial sea-level rise, or had their growth inhibited by the arrival of cold water masses. Nevertheless, the occurrence of successive phases of initiation, growth and termination of coral reefs, representing the whole period of the postglacial sea-level rise, has been demonstrated at different

Figure 70 The Lower Five Islands Delta (Bay of Fundy, Canada) is a glacio-marine delta, dated 14 to 13 ka BP, which was subsequently uplifted by glacio-isostatic movements (Photo A554, Jul. 1987)

depths offshore at Barbados (Fairbanks 1989). At Huon Peninsula, Papua New Guinea, where the local uplift rate is estimated at 1.9 mm/yr (which is small compared with the average reef growth rate of about 10 mm/yr), a 52 m drill core from the postglacial reef spanned the whole interval from 11 to 7 ka BP, showing that coral growth kept pace while the relative sea level rose by 50 m. The maximum growth rate in Huon Peninsula occurred between 10 and 9 ka BP, when the reef grew by about 20 m (i.e. 13 mm/yr after calibration) (Chappell and Polach 1991).

The development of present-day coral reefs started almost everywhere only in the Holocene (i.e. during the last 10 ka), when the rate of sea-level rise decelerated. In the northwest Caribbean, reef growth dominated by a shallow-water framework began from a shelf edge platform 20–30 m deep prior to 9 ka BP (Davies and Montaggioni 1985). In Mayotte Island, this took place 9.6 ±0.4 ka BP at a depth of about 18 m (Colonna 1994). In Mauritius, coral reefs started to develop around 9 ka BP and caught up sea level from 8.0 to 2.7 ka BP, when they approached the sea surface (Montaggioni 1988; Colonna 1994). In the nearby Réunion Island, average rates of reef growth, which were between 7.0 and 2.5 mm/yr from 8 to 5 ka BP, decreased to about 1 mm/yr during the last 5 ka (Montaggioni 1988; Colonna 1994). In the northern Great Barrier Reef region, reef growth began about 8 ka BP and reached present sea level between 5 and 2 ka BP

(Davies et al. 1985). In the Tuamotu Atolls, reef initiation appears to have occurred between 8 and 7 ka BP (Davies and Montaggioni 1985). In New Caledonia, initiation of reef growth varies according to location: the southern reefs were earlier, generally prior to 5 ka; the northern structures are younger, occurring after 4.2 ka B.P; fringing reefs reached the sea surface generally between 5 and 2.5 ka BP, after the stabilization of sea level (catch-up reefs) (Cabioch et al. 1995). Lastly, in Okierabu Island (Ryukyus), reef formation began around 7 ka BP at 11 m below present sea level. The reef was constructed by a uniform facies of *in situ* tabular corals and kept up with sea-level rise; from 7 to 6 ka BP the reef grew at a depth of 5 m; from 6 to 5.5 ka BP, as sea-level rise slowed, the reef maintained its growth rate and came to within 2.5 m of the sea level; finally, after 5.5 ka BP, the reef gradually grew to sea level (Kan et al. 1995).

At about 6 ka BP, most of the deglaciation had been completed and most present-day deltas and coral reefs were already in place. The great ice sheets of the northern hemisphere had largely disappeared, with the exception of that of Greenland. The Antarctic ice sheet might have continued to add some meltwater to the oceans, to raise sea level globally by about 2 m over the next 1500 to 1000 years (Lambeck 1993b). Any relative sea-level change after this period is a consequence of the ongoing adjustment of the earth to the redistribution of the ice load and water load that occurred during deglaciation, to (seismo)tectonic effects, and to climatic and oceanographic changes.

5.2   CONTINUANCE OF ISOSTATIC EMERGENCE/ SUBMERGENCE PROCESSES

For the near-field localities, where the ice-load effects are predominant, the continuance of glacio-isostatic processes usually results in an approximately exponentially decaying shoreline displacement curve with little, if any, change in the curvature. This is shown, for example, by the data reported from southern Norway (Fig. 71), from the western part of Hudson Bay (Fig. 72), and from Québec (Fig. 73).

Near the ice margin, relative sea-level curves become more complex. On the southwest coast of Norway, rates of change and oscillation amplitudes decrease from north to south (Fig. 74): a rapid relative sea-level fall until about 9 ka BP; a rise between 9 and 8 ka BP (continuing in some cases until 6 ka BP), when the rate of eustatic rise in sea level became greater than the local rate of isostatic uplift; and finally a gradual sea-level fall at decreasing rates of isostatic origin since 6 ka BP, which is still underway. The changing amplitude of the sea-level oscillation between 9 and 8 ka BP is probably related to the distance of the sites considered from former major ice loads. On the eastern Baltic coasts, where marks of emergence

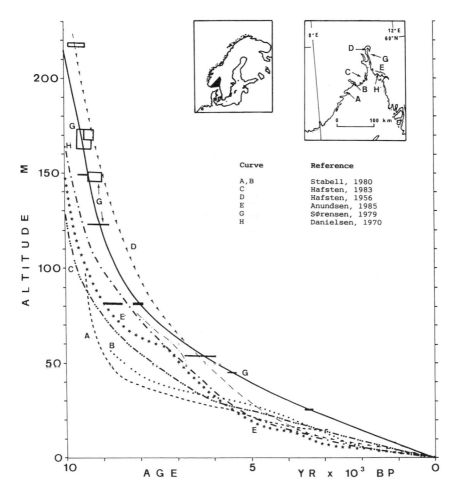

Figure 71   Holocene relative sea-level changes in the Oslofjord area, Norway. The steepness of the curves increases towards the inner part of the fjord, where former ice thickness was greater. For more details and references, see Pirazzoli (1991)

disappear in southern Lithuania and are completely absent along the coasts of Poland and Germany, relative sea-level curves show a gradual transition between uplift and subsidence (Fig. 75); some curves show important sea-level fluctuations, but these change with each author and probably result from local effects, or from poor resolution of the sea-level indicators used. A similar transition from uplift to subsidence can be observed along the Atlantic coast of Canada (Fig. 76).

For intermediate-field sites, like most of the Atlantic coasts of continental western Europe, the typical relative sea-level curve is one showing a continuous gradual rise, at rates varying with the distance from the

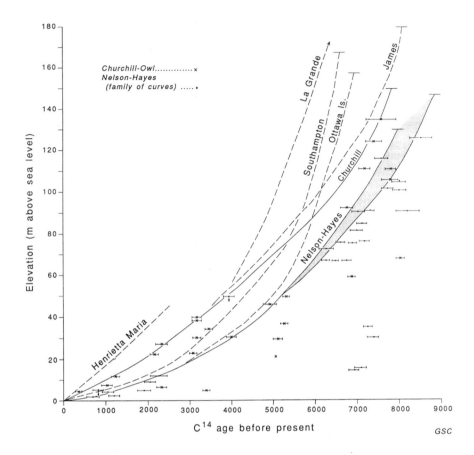

Figure 72   Emergence curves for northeastern Manitoba and other parts of Hudson
Bay, according to Dredge and Nixon (1992)

former ice load (Fig. 77). A similar situation is reported from the east
American coasts, from Massachusetts to as far as Bermuda and the West
Indies (Fig. 78).

In far-field areas, mid-Holocene emergence up to a few metres has been
reported from many coastal areas (Figs. 79–81). This emergence can be
explained, according to isostatic modellers, by the loading of the sea floor
by the meltwater in late glacial time, as a consequence of which flow is
induced in the mantle from beneath the ocean lithosphere to beneath the
continental lithosphere, producing a tilting of the continental margin and
shelf (Lambeck 1993c). The amplitude of the highstand is expected to
decrease with increasing seaward distance from the coast and to increase
inland along the shores of narrow gulfs or tidal estuaries (e.g. in the Spencer
Gulf area, southern Australia, according to Nakada and Lambeck, 1989).

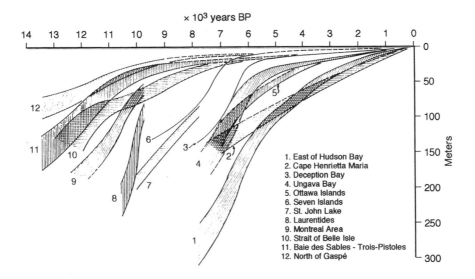

Figure 73   Emergence curves in Québec (Hillaire-Marcel and Occhietti 1980)

The amount of Holocene emergence also depends on the mantle structure. In particular, broad continental shelves with relatively thick lithosphere may impede the development of highstands along inland sites. The latter effect has been observed, for example, along the Sénégal River in West Africa and in the Alligator River area in northern Australia.

The Sénégal River estuary is developed along a very low slope perpendicular to the West African continental margin, in which the brackish water interface can be found more than 200 km upriver during the low-water stage. Holocene marine shells found *in situ* at distances of as much as 120 km from the coast (Fig. 82) have shown that the seaward tilting since 6.5 ka BP is less than about 1 m in this coastal area of West Africa, whereas an emergence of about 4 m since 5 ka BP was predicted by global isostatic models (Faure et al. 1980). In the Alligator River area, where the tidal influence extends more than 100 km upstream, the sea reached its present level before 6 ka BP and no evidence has been found for falling sea level in the last 5 ka. This is unexpected in the light of hydro-isostatic models of shelf deformation and may imply regional slow tectonic sinking (Woodroffe et al. 1987).

Mid-Holocene emergence may be much smaller in oceanic islands (Fig. 83) than along the continental margins of the far field. This is due not only to the dip-stick effect (Fig. 5), but also to the effect of the island size: small islands move up and down with the sea floor, but large islands may be subject to differential movement compared with the surrounding sea floor if some flow of mantle material of hydro-isostatic origin occurs from beneath

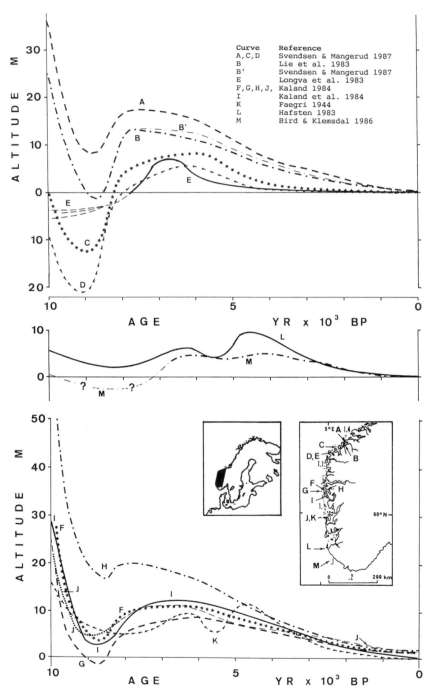

Figure 74   Holocene relative sea-level changes in southwest Norway. For more
details and references, see Pirazzoli (1991)

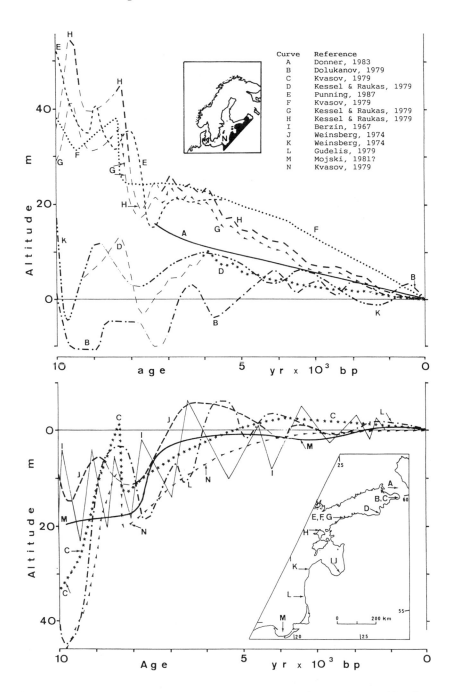

Figure 75 Holocene relative sea-level changes on the eastern Baltic shores, according to several authors. For more details and references, see Pirazzoli (1991)

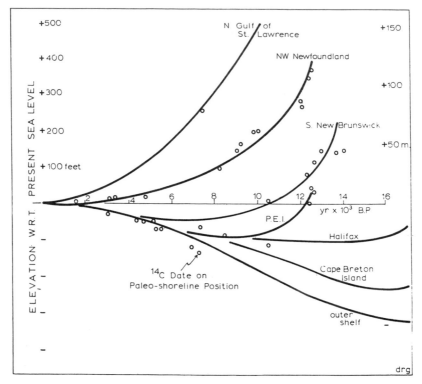

Figure 76   Local approximate variations of postglacial relative sea-level change along the Atlantic coast of Canada, illustrating the interplay of isostatic rebound, eustatic recovery and isostatic Holocene subsidence (from Grant 1980)

the ocean lithosphere to beneath the island, as beneath the continental margins (Nakada 1986; Lambeck 1993c).

Global isostatic models usually predict in the far field a well-defined highstand of sea level at about 6.0 ka BP, followed by a gradual fall to the present sea level. Such a gradual drop has indeed been verified in north Queensland, where a gradual relative sea-level fall, from about +1 m to zero, has been inferred since 6.0 ka BP (Chappell et al. 1983). The situation in French Polynesia, with a stillstand of the ocean level at about +1 m between 5 ka and 1.5 ka BP (Pirazzoli and Montaggioni 1988), is similar to that of New Caledonia (Coudray and Delibrias 1972), but different from that in north Queensland. According to Lambeck (1993c) the fact that the mid-Holocene highstand occurred later than 6 ka BP and remained at a nearly constant level for over 3 ka in French Polynesia, would support the assumption that eustatic sea levels rose globally by some 2 or 3 m during the past 6 ka, owing to some residual melting from Antarctica. The latter point has not yet been fully demonstrated, however;

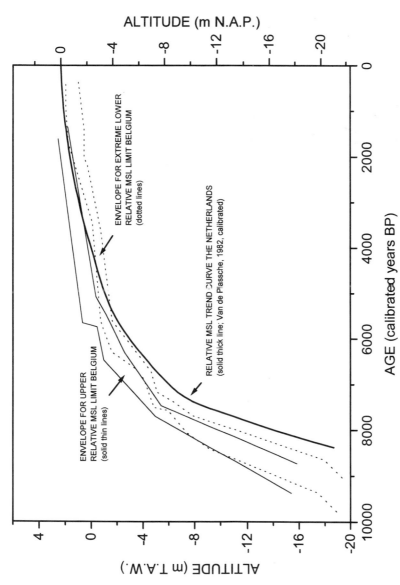

Figure 77 Comparison of Holocene relative sea-level envelopes from Belgium with the calibrated MSL curve from the western Netherlands (after Denys and Baeteman 1995). The higher position of the Belgian curve prior to 4 ka ago indicates differential crustal movement, probably of glacio-isostatic origin

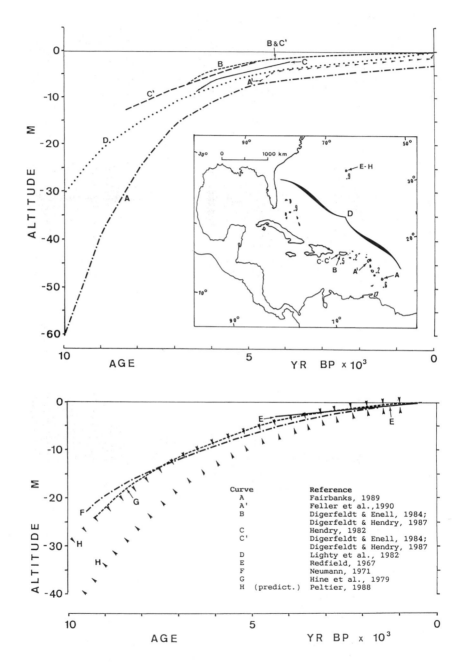

Figure 78   Holocene relative sea-level changes in the West Indies and Bermuda.
For more details and references, see Pirazzoli (1991)

Figure 79 On the right side of the Itapanhau River, near São Paulo, Brazil, black lagoonal clays dating from about 6 ka BP, visible just above the water surface, are capped by marine sands reaching an elevation of about 5.5 m above the present high-tide level on the coast, which is now several kilometres distant. These marine deposits are clear evidence of late Holocene emergence (Photo 4892, Sep. 1978)

Figure 80 Near La Plata, Argentina, the Salado River cuts through elevated marine shell beds belonging to Holocene beaches, giving clear evidence of emergence (IGCP-274 field-trip) (Photo B967, Nov. 1990)

Figure 81   Late Holocene exposed coral reefs now located near the high-tide level give evidence of 1.5 ± 0.5 m emergence near Dayyer, Iran (Photo E34, May 1994)

Figure 82   The Ferlo (Bounoum) is a former left-side tributary, now infilled, of the Sénégal River. Its terminal infilling consists of a decimetric layer containing late Holocene marine or brackish fauna (Photo 5221, Nov. 1979)

Figure 83   Remnants of a slightly elevated reef flat at Bora Bora (Society Islands, French Polynesia), containing many late Holocene corals in living position. The difference in elevation between the fossil corals and the living counterpart (*Porites* microatolls, well visible just below the water surface) is about 0.6 m (Photo 7650, Oct. 1982)

other oceanic islands do not exhibit the same duration of highstand as French Polynesian islands and relatively stable coasts of New Zealand do not show any appreciable sea-level deviation from the present situation since 6 ka BP (Gibb 1986).

5.3   SEISMO-TECTONIC DISPLACEMENTS

Site-to-site differences in height between late Holocene shorelines cannot always be explained only by glacio- and hydro-isostatic processes (Fig. 84). When shorelines are tilted over short distances, or present sudden lateral discontinuities or multiple levels (Fig. 85), or provide evidence of vertical displacements which occurred too rapidly to be consistent with gradual isostatic displacements, then seismo-tectonic processes may be involved.

During a period of relative sea-level stability, many active sea-level indicators will develop near the shoreline, capping or replacing all marks of previous stillstands. If a rapid change in the relative sea level occurs, some of these indicators may be preserved, either exposed (in the case of emergence) or concealed (often unconformably) under new sediments (in the case of submergence). When recognized in active seismic areas, these

Figure 84   Difference in height between Holocene shorelines in Tierra del Fuego.
(A) A sequence of beach ridges reaching the same elevation in San Sebastian Bay
suggest relative sea-level stability (mean spring tidal range is about 8.7 m) (Photo
B999, Nov. 1990). (B) Sequence of four elevated stepped marine terraces, reaching
about 8 m in elevation near Ushuaia (mean spring tidal range is 1.2 m). The terrace
levels have probably been displaced by coseismic uplifts, around 5.6, 4.3, 3.1 and
0.4 ka BP (Gordillo et al. 1992) (Photo D79, Nov. 1990)

Figure 85 Erosional marks of at least five stepped emerged shorelines can be distinguished on the limestone cliff between the present MSL and +3.0 m at Zambica, Rhodes Island, Greece. They have been raised by seismic events which took place during the last 6 ka (for further details, see Pirazzoli et al. 1989) (Photo 38-14A, May 1978)

indicators can be used to reconstruct and date vertical displacement movements related to major earthquakes. In some cases, even sequences of successive displacements can be identified and dated.

In many tectonically uplifting areas, the uplift trend may appear gradual in the long term but consist of a series of sudden uplifting movements interspersed with relative quiescence. Land displacements occurring spasmodically at the time of an earthquake are called *coseismic* (see Section 1.3.2) (Fig. 86). The duration of the interval between two coseismic events (*interseismic* period) is not regular, and depends on the variability of the local tectonic stress accumulation; it may tend, however, to repeat with statistical (i.e. on average) regularity. This recurrence (repeat, return) time (period) can vary, depending on the seismo-tectonic area considered, from a few centuries to over 10 000 years. Investigation and dating of former shorelines may therefore be used in active seismic areas to determine the age, distribution and succession of vertical displacements of seismic origin. When the recurrence time of morphogenetic earthquakes is relatively short, it is even possible to predict, if not when the next earthquake will occur, at least the probable areas of impending great-magnitude earthquakes and the

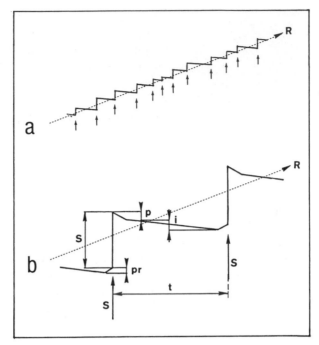

Figure 86   (a) In seismic regions where uplift is predominant, the R uplift trend may consist in the short term of sudden uplifting movements, accompanying great earthquakes, separated by more or less long periods of seismic quiescence and/or of gradual subsidence. (b) Details of movements: pr = preseismic, S = coseismic, p = postseismic, i = interseismic, t = recurrence time (from Pirazzoli et al. 1996b)

kind of vertical crustal movement that may be expected to accompany such earthquakes.

Several case studies in this field have been investigated, e.g. in New Zealand (Wellman 1967; Berryman et al. 1989; Ota et al. 1990, 1991), Japan (Yonekura 1972, 1975; Nakata et al. 1979; Yoshikawa et al. 1981; Ota 1985; Kawana and Pirazzoli 1985), Taiwan (Liew et al. 1993), the New Hebrides Islands (Taylor et al. 1980; Jouannic et al. 1982), British Columbia (Mathewes and Clague 1994), the west coasts of the USA (Plafker and Rubin 1967; Atwater 1987; Darienzo and Peterson 1990; Savage and Plafker 1991; Clarke and Carver 1992; Bucknam et al. 1992), Chile (Kaizuka et al. 1973), and Tierra del Fuego (Gordillo et al. 1992).

To demonstrate that a vertical displacement could only have taken place in a very short period of time may not be an easy task. Geomorphological indicators (stepped erosional benches or depositional platforms, ripple notches) may give useful indications but can also lead to ambiguous interpretations, since the effects of an instantaneous movement cannot always be distinguished clearly from those of events occurring in decades or

Figure 87   Marine crusts consisting of oysters, vermetids and serpulids which were uplifted coseismically at Nogimasaki, near Tokyo, at the time of the 1923 Kanto earthquake. Scale is about 1 m (Photo 2350, Dec. 1974)

even in centuries. *In situ* biological indicators can be more useful, though vertical discontinuities in elevated sublittoral populations may depend, in some cases, on factors other than rapid displacements. Anyway, to be fully convincing, the interpretation of a movement as coseismic should include clear evidence (biological, morphological, stratigraphical, archaeological, historical), or at least a cluster of converging indications, confirmed by sudden discontinuities in biological zonation and radiometric dating, suggesting that the displacement could have taken place only in a very brief period.

For example, the occurrence of *in situ* fragile sublittoral or midlittoral shells or skeletons, preserved in the supralittoral zone, is possible only after a very rapid relative sea-level change (Fig. 87). This is frequently the case, e.g. when articulated *Lithophaga* shells are found elevated, still inside their burrows, without being protected by a sandy or muddy matrix (Fig. 88).

After submergence, such evidence is most often missing, since many biological sea-level indicators are either rapidly destroyed by erosion or concealed by sedimentation, or cannot be of any use if preserved unless carried down to a depth greater than the lower limit of the vertical range of the considered species. The best way to find them *in situ* is therefore by using stratigraphic analysis on cored samples, e.g. in submerged salt marshes.

Raised reef structures (vermetid–algal rims, coral reefs), on the contrary, are more or less deeply etched by midlittoral erosion, but not destroyed.

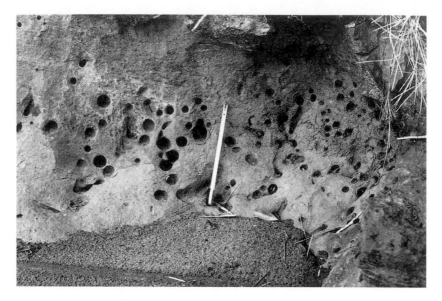

Figure 88   This mudstone formation densely bored by *Lithophaga* (some shells, articulated in living position, are still preserved inside the holes) near Himi (Toyama, Japan) emerged probably coseismically at the time of a great Holocene earthquake. Scale is about 30 cm (Photo 1679, Oct. 1974)

Their remains, though partly eroded, may still be used as sea-level markers, as has been done in many parts of the world.

The study of relative sea-level changes in seismic areas may also provide evidence of preseismic movements having preceded great-magnitude earthquakes, or, when using very recent observations, may even suggest the imminence of such a seismic event.

A well-documented case study, including repeated preseismic and coseismic movements which occurred during historical times, has been reported from a wide region of the eastern Mediterranean, extending from the Levant to central Greece. In that area, most coastal sectors showing evidence of Holocene emergence were raised by a clustering of coseismic uplifts which took place during a short period of time (approximately between the mid-4th and the mid-6th centuries AD). The regions uplifted at that time include Cephalonia and Zante in the Ionian Islands, Lechaion and the Perachora Peninsula in the Gulf of Corinth, the Pelion coast of Thessaly, Antikythira Island and the whole of western Crete, a coastal sector near Alanya in southern Turkey, the northern coast of Cyprus, and the entire Levant coast, from Hatay (Turkey) to Syria and the Lebanon. These coseismic uplifts have been demonstrated by over 30 radiocarbon dates on precise sea-level indicators. The amount of coseismic uplift was generally between 0.5 and 1.0 m but reached a maximum of about 9 m in

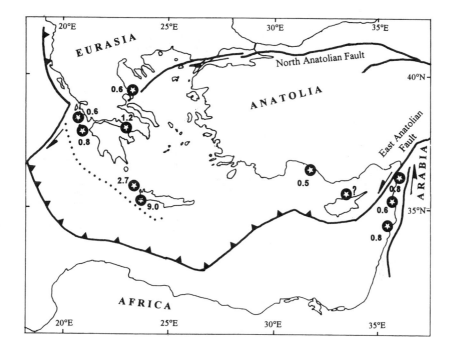

Figure 89 Location map and sketch of plate boundaries and motions in the eastern Mediterranean area. Stars: radiocarbon-dated shorelines uplifted between the middle of the 4th and the middle of the 6th century AD and amounts of uplift in metres; dotted line: active subduction trench; solid line with triangles: edge of the Mediterranean Ridge accretionary wedge (adapted from Pirazzoli et al. 1996a)

Crete (Pirazzoli et al. 1996a) (Fig. 89), where marks of Holocene emergence are widespread especially in the western part of the island, increasing gradually in altitude towards the southwest (Figs. 90, 91).

The lateral continuity of the tilted shoreline in Crete is clear evidence of a genuine tectonic effect. The uppermost marks of emergence have been dated by nine radiocarbon dates to between AD 261 and 425 (calibrated date) and their uplift has been ascribed by Pirazzoli et al. (1992) to a coseismic movement, probably accompanying the earthquake of 21 July, AD 365. The very rapid (coseismic) character of the uplift in Crete is shown by: (1) the absence of marine bioconstructions younger than the uppermost Holocene shoreline at intermediate levels between that shoreline and the present sea level; (2) the excellent state of preservation of *in situ* bioconstructions corresponding to the uppermost shoreline, and of very fragile elevated sublittoral bioconstructions at lower elevations (Fig. 92); and (3) the confirmation by radiocarbon dating that the *in situ* sublittoral bioconstructions emerging at lower levels have the same age as the uppermost shoreline, i.e. that they were uplifted at the same time.

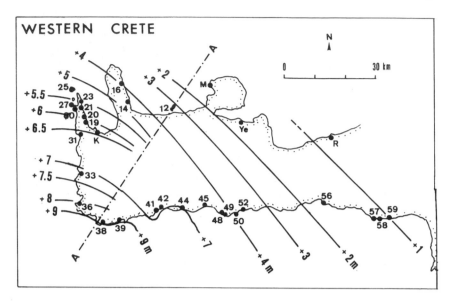

Figure 90   Elevation contours in western Crete of the shoreline uplifted coseismically probably in AD 365. Numbered dots correspond to sites of field work. A–A
indicates the average direction of the uplift gradient

Figure 91   Marks of an uplift of more than 5 m are clearly visible on the
limestone cliffs in western Crete (near Loc. 23, Fig. 90). The uplift is ascribed to
coseismic displacement associated with the earthquake of 21 July, AD 365 (Photo
3978, Oct. 1977)

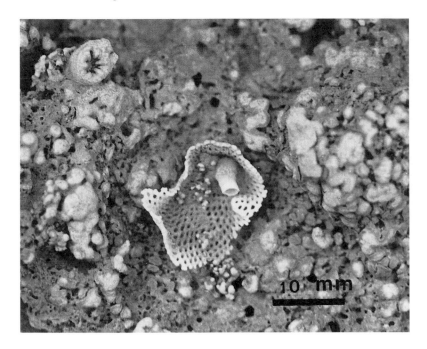

Figure 92   This reteporiform bryozoan was found at 6 m above sea level in a raised cave on the western coast of Crete. The state of preservation of its skeleton is quite perfect; such preservation would not have been possible without an extremely rapid, probably sudden emergence (from Thommeret et al. 1981; photo by J. Laborel)

In some of the areas considered in the eastern Mediterranean, the 4th to 6th century AD uplift provides the only evidence of Holocene emergence. In other areas, however, it appears to be just one of several repeated uplift movements following a longer-term trend, and the occurrence of Pleistocene shorelines has been reported at higher elevations.

The clustering of coseismic uplifts in such a wide area may be explained by the fact that the great subduction earthquake which uplifted and tilted northeastwards, probably in AD 365, a block of lithosphere approximately 200 km long, including western Crete and Antikythira, may also have produced, during the two subsequent centuries, an activation of normal fault blocks in central Greece, where rapid extension phenomena predominate, and of strike slip near the western boundaries of the North Anatolian and Dead Sea fault zones. This implies a transfer of stress all along the nearby boundaries of Anatolia–Aegea, triggering a series of other major seismic events in an area at least 1500 km across (from Syria to the Ionian Islands and to Thessaly).

In some of the areas studied in the eastern Mediterranean, certain sea-level indicators give evidence of subsidence having preceded coseismic uplift.

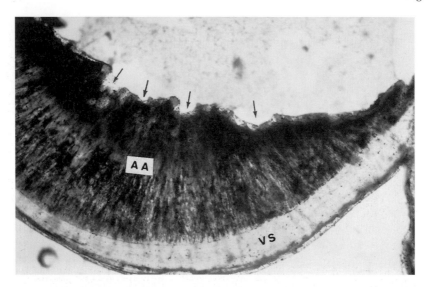

Figure 93   Petrogenic sequence in a *Dendropoma* sample collected at +2.7 m in Antikythira Island, Greece, showing that a submergence phase preceded final emergence. VS, vermetid shell; AA, precipitation of intraskeletal, epitaxial acicular aragonite; arrows indicate boring of the aragonite cement by *Cliona* (sponges) during a submergence stage (from Pirazzoli et al. 1982; photo by L.F. Montaggioni)

This is the case, for example, of Antikythira Island, where localized traces of *Cliona* borings on *Dendropoma* bioconstructions are proof of a short submergence phase just before the coseismic uplift (Fig. 93). It is also the case on the coast of Thessaly, where a coseismic uplift of 0.5 m (dated by radiocarbon to between AD 270 and 450 (calibrated date)) was preceded by a short-lived gradual preseismic subsidence of about the same amount, since borer shells in living position (which were used to date the coseismic uplift) were found on the roof of a well-developed tidal notch, whereas in normal conditions they cannot develop higher than the notch base (Fig. 94).

5.4   RELATIVE SEA-LEVEL CHANGES PRODUCED BY ASEISMIC TECTONICS

Tectonic vertical land displacements may also occur gradually, without sudden seismic jerks. This is the case of many deltaic plains, where a natural geological subsidence, frequently active over the long term, has been increased by the compaction of Holocene sediments, often accompanied by human-induced subsidence following the extraction of water, oil or gases from underground. These phenomena have been especially active

Figure 94   At Damouchari, Thessaly coast, Greece, articulated *Lithophaga* shells, which in normal conditions cannot develop higher than the floor of tidal notches, are here preserved inside their holes (arrows) near the roof of this notch (from Pirazzoli et al. 1996b; Photo D500, Aug. 1992)

in the Po Delta, Italy, where long-term natural subsidence alone can be estimated to be of the order of 1 mm/yr. The development of the modern part of the delta has taken place mostly during the last four centuries. Reclamation, involving pumping out of water, had already been accomplished by the early years of the 20th century, which resulted in the sinking of a considerable amount of land. Moreover, additional subsidence phenomena were caused by the extraction of methane-bearing waters from 1938 to 1964 (Fig. 95). Though subsequent levellings have shown a decrease in subsidence rates since 1964, when methane extraction was stopped in this area, the southern part of the delta was still affected by rapid compaction phenomena in the 1970s, with subsidence rates varying between 5 and 20 mm/yr (Bondesan et al. 1995). On the whole, the change in the relative sea level during the last century can be estimated to have varied between 1.0 and 2.7 m in the delta area, and all the land (except the dykes) is today largely below sea level.

Outside former ice-sheet areas, aseismic uplift is not unusual in volcanically or tectonically active regions. The Phlegraean Fields caldera near Naples, Italy, is especially well known for its up-and-down land movements. In the temple of Serapis at Pozzuoli, molluscan borings are developed on limestone columns up to almost 6 m above the floor of the

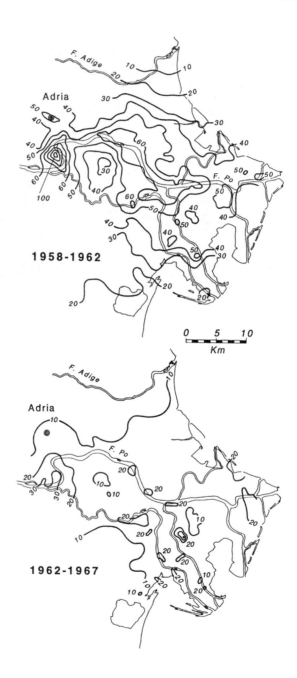

Figure 95   Contour lines showing land subsidence in the Po Delta area (in centimetres), between 1958 and 1962 and between 1962 and 1967 (from Bondesan et al. 1995, after Caputo et al. 1970)

Figure 96   Following the recent uplift in Pozzuoli, a new temporary lower passage had to be built to reach the boats in the small harbour (Photo D395, Oct. 1991)

temple, proving a period of submergence. The predominant trend here, since at least the time of construction of the temple in the 2nd century BC, was one of subsidence, at the average rate of about 10 mm/yr. The temple was therefore flooded by sea water a few centuries after its construction and subsidence reached a maximum of 12 m in the 10th century AD. The subsiding trend is, however, interrupted dramatically each time there is a lava flow beneath the caldera area. One such occurrence in the Middle Ages produced a 7 m uplift (which made the temple emerge again) and culminated in the Monte Nuovo eruption in 1538, followed by new gradual subsidence. In the 1960s the temple started to be flooded again. Two brief periods of rapid uplift occurred from 1970 to 1973 and from 1982 to 1984, fortunately without new eruptions, which would have been devastating in this densely populated area. As a consequence, a 3.2 m uplift was measured between 1968 and 1984 (Berrino et al. 1984; Pirazzoli 1995). Such a rapid uplift created several problems for coastal installations (Fig. 96).

5.5   TRANSGRESSION–REGRESSION SEQUENCES AND SEA-LEVEL CHANGES

Lithostratigraphic evidence of marine deposits interfingered with fresh-water or terrestrial sediments is classic evidence of transgression–regression sequences (overlaps). Interfingered layers are, however, rarely continuous in space and time, making the interpretation of such sequences, as well as correlation between regions in terms of sea-level changes, sometimes equivocal. As a matter of fact, such sequences are especially frequent in late Holocene sediments deposited near river mouths, but may also be found in coastal areas near major oceanic currents or upwelling phenomena, near plate boundaries where vertical coseismic displacements may have taken place, and in any coastal region affected by tsunami waves or storm surges.

A marine transgression may occur not only with a rising sea level, but even with a falling sea level if sediment supply is depleted and erosion can occur; conversely, a regression of the sea often results from a sea-level fall, but may also occur with a rising sea level in the case of high sediment supply and coastal progradation. A unit of coast may therefore display both progradation (regression) and erosion (transgression), without any movement of sea level. The author agrees with Shennan (1982a,b) and Tooley (1982) on the fact that the terms "transgression" and "regression" are totally unsuitable as formal chronostratigraphic terms and should be used only for process and lithostratigraphic description.

Among early sea-level studies, some interpretations of transgression–regression sequences on the southern coasts of the North Sea (e.g. Tavernier and Moormann 1954), considered transgressions and regressions as almost synomymous with sea-level rises and falls. Three Dunkerquian transgressions were distinguished in the upper part of the Flandrian transgression, above the Calais deposits. For each transgression a former sea-level position was deduced from the present elevation of the deposits. Correlations between the Flemish Dunkerquian stratigraphic sequences and other local deposits were subsequently attempted by various workers in France, the Netherlands and Germany, giving rise to the reconstruction of sea-level histories showing oscillations of varying amplitudes. However, 40 years later, the precise amount of these Dunkerquian sea-level oscillations, as well as their existence, remains to be demonstrated.

Near river mouths, the position of the marine–fresh water interface may depend on several factors: (a) the eustatic sea level; (b) vertical land movements (whatever their origin); (c) changes in the ocean-surface topography (which depends on oceanic currents and climate); (d) the sediment input (marine and fresh-water sediment and biomass production); (e) the local rate of sediment compaction; (f) the river flow and related salinity changes, which depend on climatic changes implying rainfall variations in the catchment basin of the river; (g) changes in storminess; and (h) changes in local wind direction, which may modify the littoral drift. The development of interfingered marine and fresh-water deposits may result from the interaction of several of the above factors.

In the North Atlantic, as shown by Ruddiman and McIntyre (1981) and Keffer et al. (1988) (Fig. 61), the northward retreat of the polar front during the last deglaciation enabled the Gulf Stream to extend northeastwards, developing the mild North Atlantic Drift and Norwegian Current. The arrival of this mild, less dense water was certainly not gradual, especially in marginal sea basins, but was episodic, with to-and-fro movements, depending on the uncertain wandering of marginal Gulf Stream meanders. The Younger Dryas, typically involving shifts in sea-surface temperature of $\geq 5°C$ in fewer than 40 years (Lehman and Keigwin 1992), can be considered as the greatest of these to-and-fro displacements during late glacial times. These led to equally large and rapid changes in atmospheric temperatures and to shifts in Atlantic deep thermo-haline circulation and ice-sheet melting rates.

At the beginning of the Holocene, most of the North Sea basin was still dry land. The British Isles became isolated from the European continent only between 8.5 and 8.0 ka BP, when wide areas of marshes and lakes were flooded by sea water and the southern North Sea basically attained its present shape (Fig. 97).

Steric and dynamic changes in sea water must have accompanied the arrival of mild North Atlantic Drift meanders round the British Isles and in the North Sea, causing small sea-level fluctuations even over a regional scale. This must have occurred several times during the Holocene and may explain at least some of the small fluctuations reported from these coastal areas.

The analysis of sea-level index points from a single site with intercalated layers of peat and clastic sediments will usually not show unequivocal fluctuations in reconstructed sea levels, owing to the errors involved in the estimation of altitude and age. Because changes in sea level affect wide areas of low coasts, "it is necessary to identify the evidence for similar and dissimilar tendencies at various sites. A rise, or fall, in sea-level will be recorded by a change in vegetation and/or lithology the nature of which is site-dependent yet the direction of change or *tendency of sea-level movement* will be the same over a much wider area" (Shennan 1982a, p. 59).

Important methodological improvements in the interpretation of transgression–regression sequences in the North Sea area were provided especially in the 1980s, in the framework of the activities of the "sea level" IGCP projects 61, 200 and 274 (see Introduction), making possible, in some cases, convincing identification of ocean-level changes as the main cause of certain transgressions and regressions. In northwest England, using 85 radiocarbon dates from three areas, Tooley (1982) recognized 12 *transgressive overlap* periods and 12 *regressive overlap* periods, though there was no single site at which all the overlaps were recorded. The consistency in the records throughout northwest England enabled Tooley (1982) to suggest an overriding process such as positive and negative sea-level movements. The above conclusions are mainly qualitative, however, i.e. based on descriptive terms, with no involvement of coastal processes.

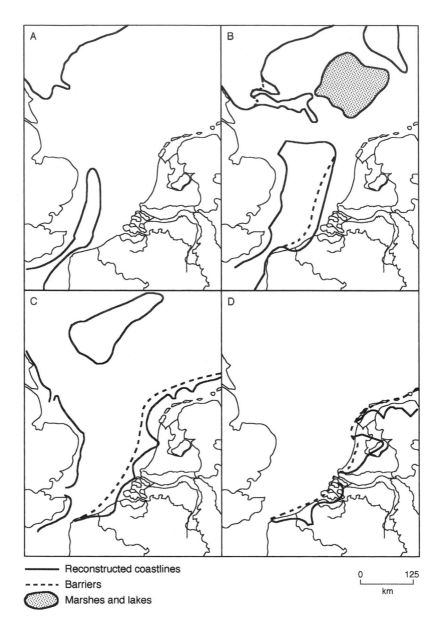

Reconstructed coastlines

- - - - - Barriers

Marshes and lakes

0   125
km

Figure 97   Reconstruction of the shoreline positions in the southern North Sea, from dates of cored basal peat, around 9 ka BP (A) 8.5 ka BP (B) 8.0 ka BP (C) and (D) 5 ka BP, according to Beets et al. (in press). The existing data suggest that the rate of relative sea-level rise decreased between 8.5 and 7.0 ka BP from NE to SW, consistently with expected glacio- and/or hydro-isostatic movements

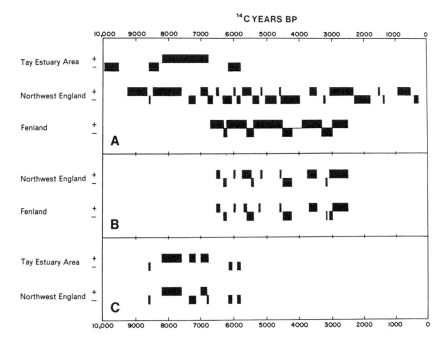

Figure 98 Regional tendencies of sea-level movement indicated by correlation of data from northwest England, the Fenland and the Tay Estuary. The chronologies of each are shown in A; the regionally significant tendencies in B and C (from Shennan, 1986)

A similar number of transgression–regression sequences have been identified by Mörner (1969, 1976) along the glacio-isostatically tilted coasts of the Kattegat; lesser sea-level oscillations have also been reported from the coasts of the Netherlands (Van de Plassche and Roep, 1989). Indeed, it would be unrealistic to expect that the same sequences are recorded everywhere in the same manner, because low-amplitude changes in sea level will not be recorded in all types of environment, and especially because dynamic sea-level factors have not remained constant on the scale of the North Sea during the Holocene (Shennan 1989). However, a steric and dynamic origin of small, regional sea-level fluctuations seems likely in areas marginal to major oceanic currents or near upwelling phenomena, where significant changes in water density may have occurred during certain periods.

The new concept of sea-level tendency introduced an objectivity to correlation schemes showing transgression sequences. An example of application is given in Fig. 98, where regionally significant correlations have been obtained by comparing dated transgressive and regressive overlaps and tendencies of sea level from rising and subsiding areas (Shennan et al. 1983).

    Application of a similar method on the shore of Connecticut enabled
Van de Plassche (1991) to infer that repeated transgressive and regressive
marsh subenvironmental changes were related to five periods of accelerated
rates of mean high-water rise during the last 2 ka, probably caused by
climate-induced changes in ocean volume and ocean mass.

    Relative sea-level fluctuations related to temporary interruptions or
reactivations of upwelling processes can be expected to have occurred in
coastal areas where such phenomena are predominant. On the coasts of
Mauritania and Senegal, where upwelling is now active, relative sea-level
curves fluctuating during the Holocene have been proposed by Einsele et al.
(1974), Faure (1980) and Barusseau et al. (1989). Among the possible
causes of the inferred fluctuations, water density changes related to steric
effects were not considered by these authors, but might, together with
climate changes, be among the possible causes of certain observed transgressions and regressions.

    Near tectonic plate boundaries, stratigraphically established transgres-
sion–regression sequences may be the result of coseismic–interseismic
vertical movements. In small coastal lagoons of northern Oregon (a region
that has not experienced substantial subduction-zone earthquakes in historical times), several buried marsh surfaces, showing sharp, non-erosional
upper contacts with tidal-flat mud deposits, indicate repeated, episodic,
abrupt subsidence by 1–1.5 m of the marsh surfaces to low intertidal levels.
In contrast, lower marsh contacts with underlying intertidal muds are
gradational, indicating gradual uplift of the order of 0.5–1.0 m and development of the marsh. These vertical tectonic movements have been
interpreted by Darienzo and Peterson (1990) as reflecting coseismic strain
release (abrupt subsidence) following interseismic strain accumulation
(gradual uplift) in the Cascadia subduction zone. The recurrence intervals
between subsidence events range in this case from a few hundred years to at
least 1000 years, and there are significant ^{14}C age overlaps with subsidence
events reported from other sites in northern Oregon and in SW Washington
for at least four burial horizons, suggesting the potential for event syn-
chronicity over regional distances of at least 200 km.

    Tsunamis can deposit sand or coarser material on coastal lowlands,
producing a transgression layer. Modern examples have been reported from
Chile, Japan, British Columbia, and other areas (for references, see Atwater
and Moore, 1992). Ancient stratigraphic examples, with an onshore sheet
of marine or estuarine sand dating to the time of a major earthquake
known or inferred to have generated a tsunami, have also been reported.
Such deposits may be very similar to those left by a flood or a storm, but
can be differentiated by the fact that a flood does not contain marine
microfossils; also, that storm surges tend to repeat in time, leaving several
interfingered sand layers, whereas tsunami events, being less frequent than
storms, most often leave a single layer of coarser deposits.

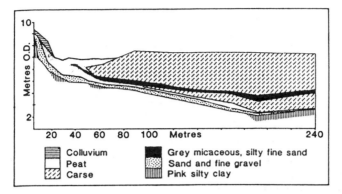

Figure 99   Profile showing lithotransgressive nature of the tsunami deposits (grey micaceaous, silty fine sand) near Fullerton, Scotland (after Smith et al. 1985, and Long et al. 1989)

Eastern Scotland, for example, was struck at about 7 ka BP by a tsunami, probably generated by underwater landsliding off SW Norway (Dawson et al. 1988; Long et al. 1989). The tsunami sediments capping former mudflats, which have been identified in several hundred boreholes, consist of a persistent and widespread layer of grey, micaceous, silty fine sand, containing abundant diatoms (Fig. 99). The run-up of this tsunami wave was estimated to be at least 4 m.

Near Seattle, Washington, USA, tsunami deposits brought by a wave which overran tidal marshes, mantling them with several centimetres of sand dated between 1.1 and 1.0 ka BP, were found at two sites. At one place, the sand sheet ascends and incorporates hillside deposits. They have been explained by Atwater and Moore (1992) as being due to a tsunami wave generated in Puget Sound, following a large shallow earthquake which occurred in the Seattle area at that time.

Recently, stratigraphical evidence was discovered of destructive tsunamis associated with the AD 1755 Lisbon earthquake, which caused extensive loss of life along the coastlines of Portugal, Spain and Morocco. The deposits are reported from Boca do Rio, Algarve, Portugal, and include laterally continuous sand layers, chaotic pebble horizons, large amounts of gravel-sized shell debris, and distinctive assemblages of marine microfossils (Dawson et al. 1995).

Frequent storm surges may produce repeated transgression layers. In temperate areas these layers are usually thin and intermittent, i.e. stratigraphically very different from the deposits which would be left by a long-lasting sea-level rise. As with tsunamis, the grain size of marine sediments left by storm surges is generally coarser than normal marine deposits in an area with similar exposure. However, storm surges do not normally produce regionally extensive deposits.

Figure 100   Limestone boulder (16×12×12 m) capping a marine terrace at +15 m, at about 150 m from the west shore of Shimoji Island, the Ryukyus, Japan. The boulder is considered to have been transported there by an undated late Holocene tsunami. The indented part of the boulder corresponds to the marine notch that had been formed along the nearby shore before the occurrence of the tsunami (after Kawana and Pirazzoli 1990; Photo 6124, Feb. 1981)

In coral reef areas, deposits left by both tropical cyclones and tsunamis may be much coarser and thicker than those left in temperate areas (Figs. 100, 101). Large blocks of coral have been left upon reefs and coastal plains of west Java and east Sumatra by the tsunami caused by the 1883 Krakatau explosion. In Funafuti Atoll, a huge rubble rampart 18 km long, with a mean height (above the reef flat) and width of 3.5 m and 37 m, respectively, was formed during tropical cyclone Bebe on 21 October 1972 (Maragos et al. 1973). The structure is commonly one-third to one-half as wide as the islets of the atoll. Rampart material originated from submarine reef slopes offshore, which were damaged to depths of 20 to 30 m; surface fragments have a mean size of about 10 cm, with storm blocks as large as 7 m. Similar rubble ramparts (though of slightly smaller size) have been observed in several Tuamotu atolls after the passage of the cyclones associated with the exceptional 1982–1983 El Niño (Pirazzoli et al. 1988b) (Fig. 102).

As a result of cyclonic waves, sand cays may be washed away and former storm ridges may migrate to leeward across reef flats. Coral debris may accumulate not only on the reef flat, but also as talus at the foot of the fore-reef slope, on submarine terraces and grooves, as lobes and wedges of debris in back-reef lagoons and as drapes of carbonate sand and mud in

Figure 101   Many boulders of coral limestone, up to several metres across, have been brought upon a marine terrace 9 m high at Hateruma Island (the Ryukyus, Japan) by a tsunami wave in AD 1771 (Ota et al. 1985) (Photo 6314, Mar. 1981)

Figure 102   In Faaite Atoll, Tuamotus, the reef flat is capped by regular cusps of coral shingle forming rampart-like ridges up to 3.5 m high, as far as the eye can see. The white colour of the outer and upper parts of the ridges indicates that most of the surficial shingle has been left very recently (Photo 7947, Oct. 1983)

deep off-reef locations. In addition to coarse deposits, other features which may aid in the recognition of former storm surges in coral reef cores include the assemblage of reef biota, its species composition and the structure of the skeletons, graded internal sediments in internal cavities, certain sequences of encrusting organisms, and submarine cement crusts over truncated reef surfaces (Scoffin 1993).

Differentiation between tsunami and storm-surge deposits is often difficult in stratigraphic sequences. The expected frequency of storm surges in the area considered and between-sites correlation may, however, be of help in interpretation: tsunamis remain a relatively rare phenomenon but can affect many coastal areas at the same time, reaching even very distant coastal regions where storm surges are unknown; conversely, storm surges may be frequent in certain areas, but their effects will always be spatially limited.

Chapter Six

Present-day sea-level trends

6.1 INTRODUCTION

Geological data may provide useful information on sea-level histories over time scales of thousands to hundreds of thousands of years. The resolution of geological data alone is insufficient to quantify relative sea-level changes over shorter periods and there is often a gap during the historical period (varying from a few decades in some remote areas to two or three thousand years around the Mediterranean and in east Asia). Although in some regions archaeological and historical data may help to fill part of this gap, most late Quaternary relative sea level curves are poorly supported by reliable sea-level indicators during recent centuries, and are therefore unable to demonstrate present-day trends.

According to oxygen isotope ratios in oceanic cores, considered in relation to astronomic cycles (Fig. 103), the present interglacial period may be nearing an end and another cold glacial phase is likely to ensue. We do not know exactly when; it may be within a few thousand years, though analysis of the insolation suggests that the present interglacial might last longer (Berger and Loutre 1996). This will be marked by a global fall in temperature and sea level. On the other hand, human activities may lead to an opposite trend. Continuous records of CO_2 content in the atmosphere of Mauna Loa (Hawaii) have shown a gradual increase from 315 ppm in 1958 to about 360 ppm in 1995 (Keeling et al. 1995), compared with 260–280 ppm in pre-industrial times (Raynaud and Barnola 1985; Lorius et al. 1988) and 180–200 ppm during the last glacial maximum (see Chapter 3). The recent increase is mainly anthropic, being due to the combustion of fossil fuels, deforestation and changing agricultural land use. In addition to CO_2, other so-called greenhouse gases, like methane, ozone, nitrous oxide and chlorofluorocarbons, show a clear trend towards increasing atmospheric concentrations during the last few decades. It may be impossible to change industrial practices sufficiently to stop the increased emission of CO_2 and most other greenhouse gases to the atmosphere. Some climatic models have indicated that the warming produced by the increased greenhouse

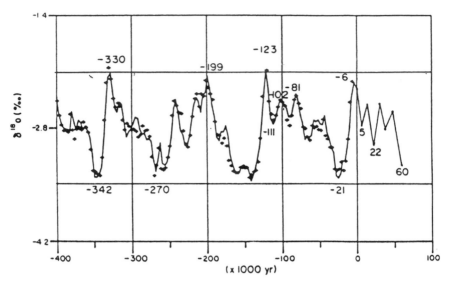

Figure 103   Long-term climatic variations over the past 400 ka and prediction for the next 60 ka according to Berger (1988). Crosses represent data from $\delta^{18}O$ deep-sea cores. The continuous line is the climate simulated by an autoregressive spectral multivariate insolation model deduced from astronomic data

effect is already in progress and has caused the sea level to rise. This would of course make extrapolation of the natural oscillations that have occurred in the past unreliable as a basis for predicting sea-level changes.

It has also been predicted that the rise in atmospheric CO_2 content will continue, if not accelerate, over the next few decades, and that, measured as CO_2 equivalent, greenhouse gases may double during the next century, reaching values which are unprecedented in the Quaternary history of the Earth. This accumulation is expected to raise the average global surface air temperature by between 1.5 and 4.5°C. This rise, which would be greater at high latitudes, is expected to produce a steric effect (the volumetric expansion of the oceans, see Section 1.4.1) and a melting of continental ice, causing a global sea-level rise.

Estimates of the sea-level rise by the year 2100 vary according to their date of publication; Fig. 104 summarizes the degree of scientific uncertainty surrounding such global predictions. In Fig. 104, heavy vertical lines are used when the estimate was actually for 2100; light vertical lines are used where the estimate was for a year before 2100 and has been extrapolated to 2100; the letter "B" indicates that the author did not estimate a range (high to low).

Since the extravagant predictions of the early 1980s, when the US Environmental Protection Agency was expecting a sea-level rise in the year 2100 of between 56 and 345 cm (Hoffman et al. 1983), there has been,

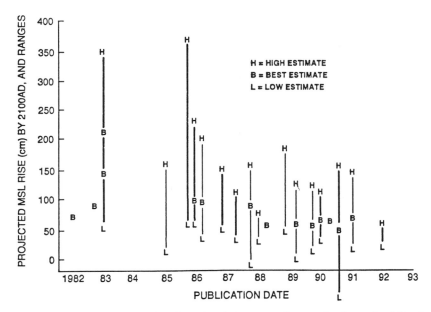

Figure 104   Predictions of global sea-level rise according to the date of publication
(after Maul et al. 1996)

during the last decade, a decline in the sea-level rise predicted for the next century. Even in 1986 a sea-level rise greater than 2 m by the year 2100 was considered possible by several authors. More recent predictions, on the other hand, consider a rise of less than 1 m more likely over the same period. According to the Intergovernmental Panel on Climate Change assessment (IPCC 1990), the "best estimate" scenario for the year 2100 was a global sea-level rise of 66 cm, with high and low estimates of 110 cm and 31 cm, respectively. A later revision (Wigley and Raper 1992) proposed a global sea-level rise of 48 cm, with high and low estimates of 90 cm and 15 cm, respectively, by the year 2100.

In this chapter, recent trends deduced from instrumental measurements will be summarized briefly. Some additional considerations will help to identify which coastal areas are expected to be more at risk of flooding during the next century.

6.2   INSTRUMENTAL MEASUREMENTS

Regular instrumental records are available only for a very recent period, often for only a few decades, but in some cases for over a century. Extremely rare records go back to the 18th century. These are often difficult to interpret because the methods, accuracy and frequency of measurement have changed several times. The main methods of instrumental measurement of sea level use tide-gauge stations (often linked to each other by national geodetic levelling networks), oceanographic devices and satellite sensing.

6.2.1   Tide-gauge records

After the first tide levels were established (in 1682 in Amsterdam (Van Veen 1954), 1704 in Stockholm (Mörner 1979), 1807 in Brest, 1811 in Swinoujscie, and 1825 in Venice), early measurements were made by reading the level of the water on a graduated staff, once or several times each day. By the second half of the 19th century more than 100 tide-gauge stations had been established in Europe (especially along the Baltic coasts), in North America, and in a few harbours in other continents. During the 20th century over 1000 tide gauges have been in use, some for considerable periods. Continuous recording devices have used ink pens coupled to a float and a clockwork mechanism and, more recently, automatic records fed directly into a computer. Monthly and yearly means from most tide-recording stations are stored in the database of the Permanent Service for Mean Sea Level (PSMSL) at Birkenhead, UK.

The initial aim of tide-gauge measurements was to serve navigation rather than science. Tide gauges were therefore generally associated with the operation of a major port and consequently are often to be found in a river or estuary, where water density may change with river runoff and sediment compaction may cause subsidence. Tide gauges may be affected by shipping movements in and out of harbours, or may be damaged or displaced by collisions.

Over given periods, relative sea-level variations can appear to be periodic (tidal), random (meteorological and hydrological) or monotonic (neotectonics, eustasy). If periodic, short-lived and random components are removed by filtering processes, and if the series investigated is long enough, long-term trends will appear. However, tide gauges measure local sea level relative to a benchmark on land, so that vertical crustal movement is one of the factors affecting the local long-term signal.

According to Warrick and Oerlemans (1990, p. 266) "accurate trends can be computed only given 15–20 years of data", but such a record would be too short in many cases. Pirazzoli (1986b) analysed most of the data stored by the PSMSL and other sources, and considered a more realistic lower limit of 50 years of data, whereas Douglas (1991) chose 60 years; according to Baker et al. (1995), records spanning at least 40 years are required for Mediterranean stations in order to have errors of less than 0.5 mm/yr. There are comparatively few tide-gauge records of such duration.

Several analyses of the trends shown by tide-gauge records and their spatial distribution can be found in the literature, the most complete being that published by Emery and Aubrey (1991). In order to illustrate the main sea-level tendencies shown by tide-gauge records, some regional graphs, taken from earlier publications (Pirazzoli 1986b, 1989), will be discussed. Most of the longest records available are from either side of the North Atlantic (Fig. 105). Their rates of secular change are shown in Fig. 106. On the American side, with the exception of station 1 (Pointe-au-Père) which is situated in an area of glacio-isostatic uplift, most tide gauges indicate a long-term relative sea-level rise, at rates varying from about 2 to 4 mm/yr and reaching as much as 6.3 mm/yr at Galveston, where anthropic effects (pumping of underground fluids) are predominant. A submergence trend along these coasts, which are situated in intermediate-field sites, has been predicted by global isostatic models. However, the rise observed is two to three times higher than the rates deduced from geological evidence in the late Holocene, and oceanic effects (steric and/or dynamic) are thought to have contributed to the recent trends found on North American coasts.

On European coasts the overall pattern is much like that foreseen by global isostatic models: a dome-shaped uplift centred in the Gulf of Bothnia, surrounded by a wide belt of moderate subsidence. If those stations indicating a relative sea-level rise in Fig. 106 (2, 7 to 25, 28, 45 to 58) are considered separately, they indicate an average rate of rise of 1.26 mm/yr (this rate will be discussed below).

The average regional relative sea-level changes prior to 1980 are shown in Fig. 107. The changes clearly vary from one area to the other, and no general trend predominates. Indeed, determination of a present eustatic trend from tide-gauge records alone remains at this stage a challenging problem, even in a part of the world where the network of stations is most dense and the longest records are available.

In east Asia, most tide-gauge records of longer than 50 years come from Japan, where sea-level trends are irregular and change from place to place (Fig. 108). The contrast is striking between the apparent relative stability of Tonoura (13), Wajima (14), Oshoro (16) and Hosojima (29) on the one hand, and the high rates of relative sea-level rise in Mera (23), Aburatsubo (24) and especially Toba (26), where it reaches 23 mm/yr. Indeed, plate tectonics has had a very strong influence in this area, long-term trends in sea level being affected by interseismic crustal deformation and by co-seismic movements at the time of earthquakes of great magnitude. This is the case of Kushimoto (27), where a rapid relative sea-level rise was interrupted by the 1944 "Tonakai" and 1946 "Nankaido" earthquakes. In Aburatsubo (24) the tide-gauge station has recorded vertical deformation: a gradual sea-level rise at the rate of about 5.5 mm/yr from 1895 to 1923 was followed by a sudden uplift of 1.37 m at the time of the 1923 "Kanto" earthquake (see also Fig. 87), and then a gradual sea-level rise at the rate of 4 mm/yr from 1923 to 1980 (only the latter part is shown in Fig. 108). The sea-level trends in Fig. 108 have probably also been affected by steric

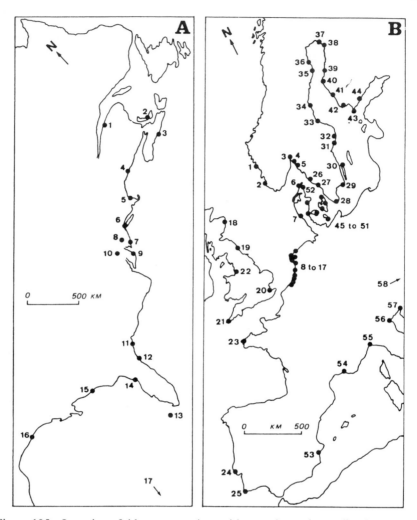

Figure 105   Location of tide-gauge stations with records starting earlier than 1925 on both sides of the North Atlantic. (A) East coast of North America: 1, Pointe-au-Père; 2, Charlottetown; 3, Halifax; 4, Portland; 5, Boston; 6, New York; 7, Atlantic City; 8, Philadelphia; 9, Lewes; 10, Baltimore; 11, Charleston; 12, Fernandina; 13, Key West; 14, Cedar Keys; 15, Pensacola; 16, Galveston; 17, Cristobal. (B) Coasts of Europe: 1, Bergen; 2, Stavanger; 3, Oslo; 4, Strömstad; 5, Smögen; 6, Hirsthals; 7, Esbjerg; 8, Delfzijl; 9, Terschelling; 10, Harlingen; 11, Den Helder; 12, Ijmuiden; 13, Hoek van Holland; 14, Hellevoetsluis; 15, Brouwershaven; 16, Zierikzee; 17, Vlissingen; 18, Aberdeen II; 19, North Shields; 20, Sheerness; 21, Newlyn; 22, Liverpool; 23, Brest; 24, Cascais; 25, Lagos; 26, Göteborg; 27, Varberg; 28, Ystad; 29, Kungholmsfort; 30, Olands Norra Udde; 31, Landsort; 32, Stockholm; 33, Björn; 34, Draghällan; 35, Ratan; 36, Furuögrund; 37, Kemi; 38, Oulu; 39, Jacobstad; 40, Vaasa; 41, Mäntyluoto; 42, Turku; 43, Hangö; 44, Helsinki; 45, Gedser; 46, Köbenhavn; 47, Hornbaek; 48, Korsör; 49, Slipshavn; 50, Fredericia; 51, Aarhus; 52, Frederikshavn; 53, Alicante; 54, Marseille; 55, Genova; 56, Venezia; 57, Trieste; 58, Port Tuapse

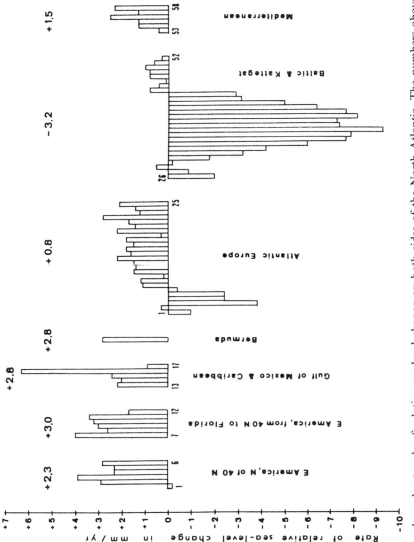

Figure 106 Local secular trends of relative sea-level change on both sides of the North Atlantic. The numbers above the graphs are average rates of change in each region (+ = relative sea-level rise; − = relative sea-level drop). The numbers near the zero line in the graphs correspond to the location numbers in Fig. 105 (from Pirazzoli 1989)

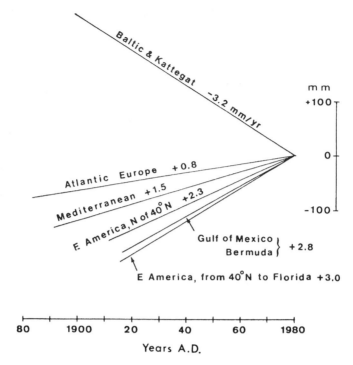

Figure 107   Average relative sea-level changes during the 100 years to 1980 on both
sides of the North Atlantic (from Pirazzoli 1989)

changes, which cause fluctuations of the order of some centimetres from
year to year and from decade to decade (White et al. 1979), in relation to
the Kuroshio current. This is not surprising, as differences as high as 2.5 m
in the topography of the surface of the ocean have been reported (Ganeko
1983), and several large-scale meanderings of the Kuroshio current have
been recorded in the past few decades.

    Along the west coast of the Americas (Fig. 109), crustal uplift has been
very strong in the Gulf of Alaska, reaching as much as 20 mm/yr at
Skagway (6). To the south, relative sea-level trends are irregular, with drops
in some places and rises in others. South of California, records are not long
enough to permit deduction of long-term trends, except in Balboa. The
highest rates of relative sea-level rise have been reported from Seattle
(+1.9 mm/yr), San Diego (+1.9 mm/yr) and Balboa (+1.8 mm/yr), whereas
lower rates have been observed at San Francisco (+1.2 mm/yr), Los
Angeles (+0.6 mm/yr) and Victoria (+0.5 mm/yr), with little change at
Vancouver (−0.1 mm/yr). However, as the west coasts of the Americas are
located near active plate boundaries, they have probably been affected by
tectonic movements. The most recent coseismic movement has occurred in

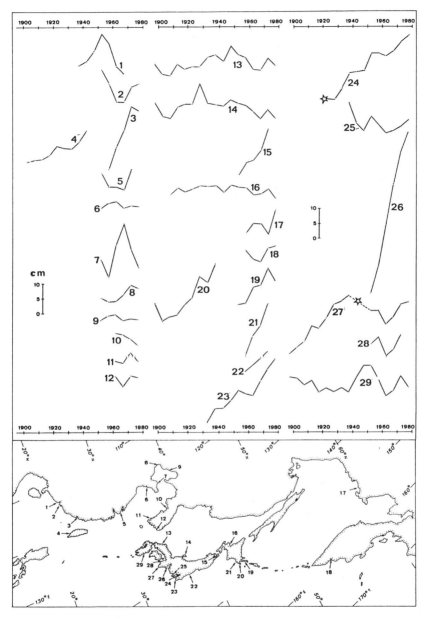

Figure 108   Average 5 year relative sea-level variations in east Asia. 1, Macao; 2, Hong Kong; 3, Xiamen; 4, Takao; 5, Shanghai; 6, Qingdao; 7, Yantai; 8, Tanghu; 9, Qinghuangdao; 10, Inchon; 11, Mokpo; 12, Pusan; 13, Tonoura; 14, Wajima; 15, Ominato; 16, Oshoro; 17, Nagajeva Bay; 18, Petropavlosk; 19, Yuzno Kurilsk; 20, Hanasaki; 21, Kushiro; 22, Onohama; 23, Mera; 24, Aburatsubo; 25, Uchiura; 26, Toba; 27, Kushimoto; 28, Kainan; 29, Hosojima. Stars indicate the occurrence of great earthquakes (from Pirazzoli 1986b)

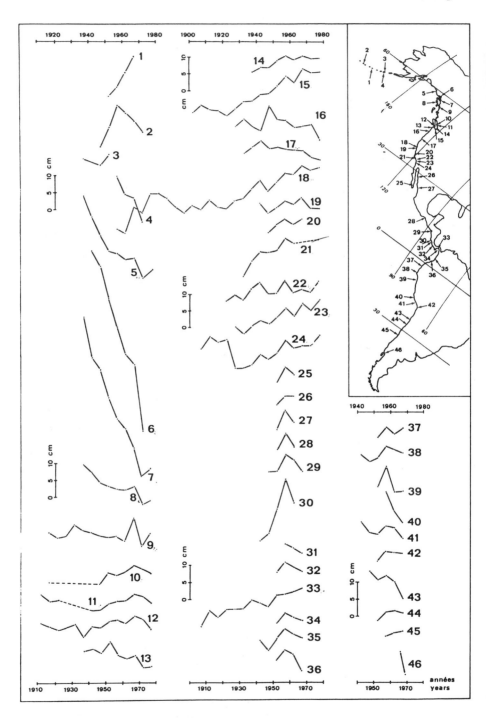

Antofagasta, where a coseismic uplift reaching a maximum of 40 cm accompanied the earthquake of 30 July 1995 (Ortlieb et al. 1995).

In spite of the limited number of sufficiently long tide-gauge records available, their very uneven geographical distribution and the difficulty of separating the various components of relative sea-level change, several authors have attempted an estimation of the recent global (average) sea-level rise (Table 4). Three main approaches can be distinguished.

The first approach, common to most early studies (1–11), consisted mainly in using various kinds of averaging methods applied to uncorrected data, after exclusion of records considered doubtful or coming from areas regarded as subject to uplift or subsidence, but not excluding wide areas which are now known to be of moderate subsidence of glacio-isostatic origin (e.g. western Europe and the Atlantic and Gulf coasts of the USA). This attempt, which obtained average values of sea-level rise ranging mainly from 1.0 to 1.5 mm/yr, is considered to be biased towards a rise slightly greater than the eustatic one (Pirazzoli 1986b).

A second approach consisted in "correcting" the primary data, using geological data (9,13,20), climatic correlations (15) or geophysical isostatic models (14,17,18). This provided more variable results, though with greater values of sea-level rise when stations from the east coast of North America (where the rate of relative sea-level rise is generally much faster than in other coastal areas of the Atlantic) are predominant in the sample of tide-gauge stations considered.

The relatively close agreement (between 0.85 and 1.2 mm/yr) obtained by independent analyses applying a geological correction (9, 13, 20) is worth noting. According to Mörner (1973), a rapid eustatic rise started about 1840, slowed down about 1930 and ended about 1950.

Geophysical isostatic models have provided estimates of land movements produced by deglaciation. They can be used, therefore, to predict present-day movements, in order to filter from tide-gauge records the present-day glacio-isostatic and hydro-isostatic components. Such a filtering has been proposed by Peltier and Tushingham (1989), applying a model similar to that used to obtain Figs. 53, 54 and 56, but the rates obtained (a global sea-level rise of 2.4 mm/yr) are surely excessive for recent sea-level rise. If

---

Figure 109 (*opposite*) Average 5 year relative sea-level variations on the west coasts of the Americas. 1, Massacre Bay; 2, Sweeper Cove; 3, Dutch Harbor; 4, Unalaska; 5, Yakutat; 6, Skagway; 7, Juneau; 8, Sitka; 9, Ketchikan; 10, Point Atkinson; 11, Vancouver; 12, Victoria; 13, Neah Bay; 14, Friday Harbor; 15, Seattle; 16, Astoria; 17, Crescent City; 18, San Francisco; 19, Alameda; 20, Avila; 21, Santa Monica; 22, Los Angeles; 23, La Jolla; 24, San Diego; 25, La Paz; 26, Guaymas; 27, Mazatlan; 28, Salina Cruz; 29, La Union; 30, Puntarenas; 31, Quepos; 32, Puerto Armuelles; 33, Balboa; 34, Naos Island; 35, Buenaventura; 36, Tumaco; 37, La Libertad; 38, Talara; 39, Chimbote; 40, San Juan; 41, Matarani; 42, Arica; 43, Antofagasta; 44, Caldera; 45, Valparaiso; 46, Puerto Montt (from Pirazzoli 1986b)

Table 4   Estimates of global (average) sea-level rise from tide-gauge records

Authors	Number of stations	Period of time considered	Average rate of sea-level rise (mm/yr)
1 Gutenberg (1941)	69	1807–1937	1.1
2 Polli (1952)	110	1871–1940	1.1
3 Cailleux (1952)	76	1885–1951	1.3
4 Valentin (1952)	253	1807–1947	1.1
5 Lisitzin (1958)	6	1807–1943	1.1
6 Fairbridge & Krebs (1962)	unspecified	1860–1960	1.2
7 Kalinin & Klige (1978)	126	1900–1964	1.5
8 Emery (1980)	247	1850–1978	3.0
9 Gornitz et al. (1982)	193	1880–1980	1.2
10 Barnett (1983)	9	1903–1969	1.5
11 Barnett (1984)	152	1881–1980	1.4
		1930–1980	2.3
12 Pirazzoli (1986b)	229	1807–1984	indeterminable
13 Gornitz & Lebedeff (1987)	130	1880–1982	0.9–1.2
14 Peltier & Tushingham (1989)	40	1920–1970	2.4
15 Pirazzoli (1989)	58 (Europe)	1880–1980	0.52
16 Stewart (1989)	152	1881–1980	indeterminable
17 Trupin & Wahr (1990)	84 (N of 30°N)	1900–1979	1.75
18 Douglas (1991)	21	1880–1980	1.8
19 Emery & Aubrey (1991)	517	1807–1986	indeterminable
20 Shennan & Woodworth (1992)	33 (UK & North Sea)	1901–1988	1.0±0.15
21 Gröger & Plag (1993)	854	1807–1992	indeterminable

true, such a rate would mean that subsiding areas like the Netherlands, where the rate of recent relative sea-level rise is usually less than 2.4 mm/yr, would in fact be emerging. Also, the Venice area (Italy) would be emerging, in spite of human-induced land subsidence, which has been demonstrated during the last century by repeated surveys (Bondesan et al. 1995). Although a "global" rate of sea-level change is not expected to be the same everywhere, such a high rate of sea-level rise cannot be accepted. The 1.75–1.8 mm/yr rates proposed by Trupin and Whar (1990) and Douglas (1991), using the Peltier and Tushingham (1989) model, also seem too high. The idea of using appropriate isostatic models to filter the isostatic component from tide-gauge records is useful, but in the absence of filtering of other tectonic and oceanographic processes, the results cannot be considered as purely eustatic. Unfortunately, the isostatic models presently available do not take account of ocean dynamics or tectonic effects other than glacio-isostatic and hydro-isostatic effects (Table 5). Global geophysical models are certainly useful, but are not yet complete and precise enough to provide an accurate estimate of the present-day global sea-level trend from tide-gauge records.

Table 5   Advantages and drawbacks of methods for calculating the global eustatic factor

Method	Remarks
(1) Averaging tide-gauge data (most authors in Table 4)	(1) Uneven geographical distribution of data; (2) the assumption that single stations can represent average neotectonic movements in wide regions is unacceptable; (3) the assumption that crustal movements can be compensated by using a sufficiently large number of stations biases the average towards an apparent sea-level rise.
(2) Subtracting geological trends from tide-gauge data (Gornitz et al. 1982; Gornitz and Lebedeff 1987; Shennan and Woodworth 1992)	(1) Can provide good results locally; (2) few geological sea-level data are usually available in the same area as a tide-gauge station; (3) inappropriate in active tectonic areas; (4) ocean dynamics effects are disregarded.
(3) Applying a correction factor deduced from global isostatic models (Peltier and Tushingham 1989; Trupin and Wahr 1990; Douglas 1991)	(1) Applicable to all sites; (2) simplifying assumptions in the model may cause systematic deviations from field data; (3) tectonic trends other than glacio-isostatic ones are disregarded; (4) ocean dynamics effects are disregarded.
(4) Computing steric sea-level trends with oceanographic data	(1) Short records, strong noise; (2) glacio-eustatic factors are disregarded.
(5) Subtracting the long-term trends from tide-gauge data and adding a eustatic factor estimated from comparisons between temperature changes and very long tide-gauge records (Pirazzoli 1989)	(1) Land movement and long-term ocean dynamics effects are removed; (2) the eustatic factor can be estimated only in very few sites.
(6) Using satellite altimetry	(1) This method is the only one likely to provide an absolute datum; (2) the accuracy of altimetric measurements should be adequate.

The third group of studies (12,16,19,21 in Table 4) did not succeed in assigning any reasonable estimate to the present rate of eustatic rise of sea level, because of background noise (due to meteorological, hydrological, tectonic and anthropic factors) and the limitations of the data used, which prevented accurate global assessments.

Only 13% of the 229 stations considered by Pirazzoli (1986b) indicate a rise of between 1.0 and 1.5 mm/yr, and only 17% a rise of between 1.5 and

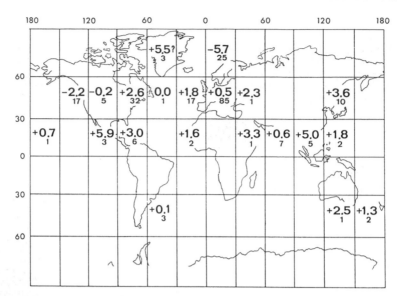

Figure 110 Distribution of average long-term linear trends of relative sea-level change indicated by tide-gauge records of longer than 50 years (and a few records 30 to 50 years long showing small variability), according to Pirazzoli (1986b). Large numbers on the upper line indicate average values (in millimetres per year; + = relative sea-level rise) deduced from the trends located in each 30° latitude by 30° longitude compartment. The small numbers on the lower line correspond to the number of stations used in each compartment

2.4 mm/yr (corresponding to values often assumed as "eustatic"). On the other hand, 21% of the stations show a relative sea-level rise of over 2.4 mm/yr, 20.5% a rise between 0.1 and 1.0 mm/yr, 1% a relative stability, and 27.5% a drop in the relative sea level. It is true that the general average corresponds to a sea-level rise of about 1 mm/yr, but the distribution of values considered departs from a typical Gaussian distribution, and one may wonder about the meaning of such an average, which is influenced by so many different effects.

The geographically uneven distribution and diversity of long-term trends in relative sea-level changes is illustrated in Fig. 110, where the Earth's surface is divided into 72 compartments by a grid with intervals of 30° in latitude and 30° in longitude. Most of the 229 stations investigated (70%) were located in only four compartments; on the other hand, no data existed in 70% of the compartments. The average trend obtained differed greatly from one compartment to another and no general trend appeared, although positive values (sea-level rise) are more frequent than negative values (sea-level fall). This diversity confirms that local and regional factors, either tectonic movements (*sensu lato*) or oceanic changes, are much larger than average global eustatic factors.

Emery and Aubrey (1991) grouped 517 tide-gauge records according to geological processes or human interactions that appear to have controlled the vertical movements of land. They found a very wide range of variation in all groups, which prevented a convincing eustatic assessment. The median displacement rate of all 517 stations considered is a relative sea-level rise of 1.1 mm/yr, but as subsidence is more frequent than uplift in coastal areas, for geological reasons (Pirazzoli 1986b; Stewart 1989), this can also be interpreted as an average rate of crustal movements and hydrodynamic changes, though it may include a eustatic component of unknown amount. That is why Emery and Aubrey (1991, p. 160) concluded "we must state that we are unable to assign any reasonable estimate to the present rate of eustatic rise of sea level" and that the eustatic "rise can be bracketed only as ranging between 0 and 3 mm/yr".

Nevertheless, it has been shown (see above in this section) that all European tide-gauge stations which are not being uplifted show an average rate of relative sea-level rise of 1.26 mm/yr. Since, for rheological reasons, subsidence must exist in at least part of this area, to balance the uplift in Fennoscandia (see Fig. 106), it can be concluded that the recent eustatic rise in Europe is certainly less than 1.26 mm/yr.

In spite of the above divergent interpretations, there are two important points on which general agreement is emerging: first, nobody claims that the global sea level has been dropping during the last 100 years; and second, in spite of increasing concentrations of greenhouse gases in the atmosphere, no evidence can yet be found for an acceleration of mean sea-level rise (e.g. Woodworth 1990; Douglas 1992).

6.2.2 Steric measurements

Atmospheric and oceanic dynamical processes account for a large fraction of the interannual sea-level variability. This variability is unfortunately greater in areas where many tide-gauge stations with long records are located (Fig. 111). Such variability, of several centimetres, introduces much background noise, which makes it difficult to identify long-term trends of the order of 1 mm/yr. Background noise is especially loud near major oceanic currents. In the Gulf Stream area, for instance, an altimetric variability of as much as 40 cm has been observed between 1975 and 1978 from repeated pairs of GEOS 3 altimeter profiles (Fig. 112). Slow lateral displacements of this variability, e.g. near the continental shelf of the USA, would seriously affect the long-term trends shown by tide-gauge records on the coast. This may help to explain why the apparent rates of sea-level rise along the Atlantic coast of the USA recorded since the 1920s are not only faster than in most other regions, but also an order of magnitude greater than the geological trends deduced in the same areas from Holocene sea-level data.

Some attempts to measure sea-level changes caused by steric effects have

150

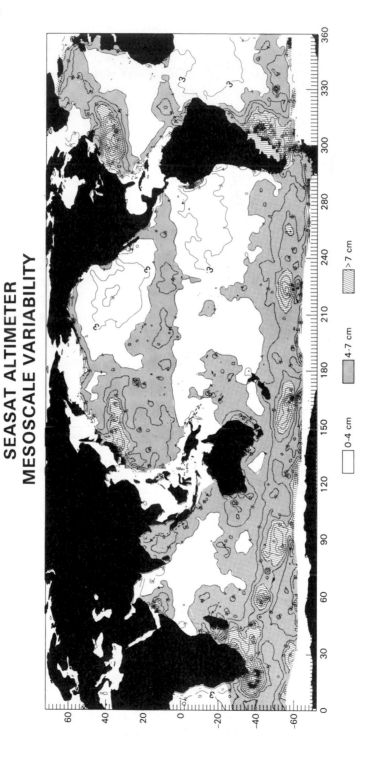

Figure 111   Global mesoscale sea height variability measured by the SEASAT altimeter from 15 September to 10 October 1978 (after Cheney et al. 1983)

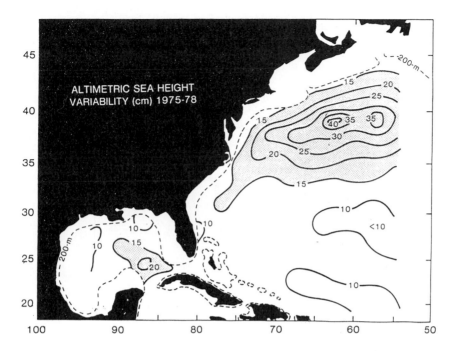

Figure 112 Sea surface mesoscale variability in the northwest Atlantic and Gulf of Mexico according to Douglas et al. (1983). Maximum values are associated with the Loop Current in the Gulf of Mexico and in the Gulf Stream meander region downstream of Cape Hatteras

proved interesting, though still inconclusive. In the northeast Pacific, a 27 year time series of monthly hydrographic observations has been studied for station PAPA (50°N, 145°W) (Thomson and Tabata 1987). Linear trends based on this record suggest that steric heights relative to 1000 decibar surface are increasing at a rate of 0.93 mm/yr and that two-thirds of this increase is due to thermosteric changes at depths below 100 m. However, a critical examination of the results indicates that sea-level changes of such small magnitude would be masked by the large (1–10 cm) interannual variability of open ocean steric height. Thomson and Tabata (1987) concluded that their trend estimates are still open to question, and that a 27 year time series is too short to permit accurate resolution of possible climate-induced changes in global sea level.

In the North Atlantic, temperature measurements were repeated along the same transect (about 24°N) in 1957, 1981 and 1992 (Roemmich and Wunsch 1984; Parrilla et al. 1994). Over the whole period from 1957 to 1992, the sea along this transect has warmed between depths of 700 and 2500 m, maximum warming (up to 0.32°C over 35 years) occurring at a depth of about 1100 m (Fig. 113). The water below 3000 m cooled, however, though at an

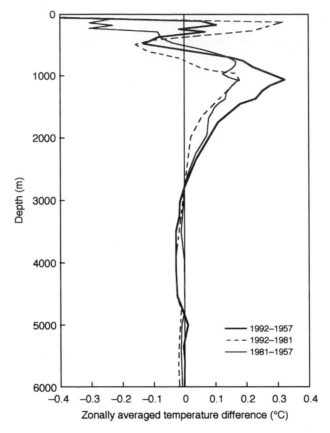

Figure 113   Vertical profiles of the zonally averaged temperature differences over the periods 1981–1957, 1992–1981 and 1992–1957 in the subtropical North Atlantic Ocean at 24°N, according to Parrilla et al. (1994)

order of magnitude less than the warming at 1100 m, for all three time periods. The strongest cooling (0.03°C) appeared at 3750 m depth, especially on the western boundary, where the southward flow of North Atlantic deep water (from its formation in the Greenland and Labrador Seas) is renewing the deep water reservoir. The warming is statistically significant between 800 and 2250 m depth and the change in dynamic height over this depth interval, which is the effective sea-level rise due to warming, is three dynamic centimetres. Nevertheless, the observed warming is surprisingly deep in the interior of the ocean in comparison with the intensified surface warming of ocean climate models.

In the Pacific Ocean, on full-depth sections between Australia and New Zealand between 1967 and 1989, there has been a depth-averaged warming below the mixed layer of 0.04°C at 43°S and of 0.03°C at 28°S. Accordingly, the sea-level rise caused by expansion between a depth of 300 m and the ocean floor is 2–3 cm (Bindoff and Church 1992).

Off the coast of southern California, temperatures have increased by 0.8°C uniformly in the upper 100 m and have risen significantly to a depth of about 300 m between 1950 and 1992. This corresponds to an expansion of the water column causing a sea-level rise of 0.9 ±0.2 mm/yr (Roemmich 1992).

On the contrary, in 1991 the waters between Greenland and the British Isles were 0.08°C colder than in 1962 and 0.15°C colder than in 1981, the cause appearing to be renewed formation of intermediate water in the Labrador Sea from a cooler and fresher source, and the spreading of this water from the west (Read and Gould 1992).

Very few datasets are yet available, and the interannual and interdecadal variability of the oceans makes problematic any application of these geographically limited observations to long-term changes in sea level.

6.2.3 Geodetic measurements

6.2.3.1 Land and oceanographic levelling

When several countries started establishing levelling networks in the second half of the 19th century, each chose a local sea-level reference. In Britain, the first primary network of levelling was made between 1840 and 1860, to a datum which approximated mean sea level at Liverpool. When national relevelling began in 1912, three gauges were established, one at Dunbar in Scotland, one at Felixstowe on the east coast and one at Newlyn in the extreme southwest (Pugh 1987). Levelling measurements made between these stations showed that mean sea level was not the same everywhere, for example, it was 25 cm higher at Dunbar than at Newlyn. Similar slopes on the mean sea-level surface from south to north have been found in the American surveys, in Japan and in the Australian network.

In the United States, surveys showed consistently higher mean sea levels on the Pacific coast than on the Atlantic coast, with a maximum apparent mean sea-level variation, relative to Galveston, Texas, of +0.59 m at Fort Stevens, Oregon, and –0.28 at Old Point Comfort, Virginia.

For comparisons between levellings in continental Europe, the Normal Amsterdam Peil (NAP, i.e. the mean level at which the Zuiderzee water could formerly penetrate into the canals of Amsterdam) was chosen as a reference level by the International Association for Geodesy. This level is 52 cm lower than mean oceanic height (Doodson and Warburg 1941). According to European levellings, mean sea level decreases about 60 cm from the Baltic to the northern Mediterranean, and 88 cm from the Baltic to the Black Sea. From the Mediterranean to the Red Sea, on the other hand, mean sea level increases by 25 cm (Lisitzin 1965).

These demonstrated differences in mean sea level correspond with permanent gradients on the sea surface due to currents, density changes, atmospheric pressure and wind, which in combination can cause mean sea level to vary from an undisturbed geoid by more than 1 m (Fig. 8). Mean sea level can only be measured and related to land levels along the coast, but the geoid is conceived as a continuous, completely closed surface, passing beneath the continents. As noted by Pugh (1987, p. 331) "to determine a true mean sea-level, measurements must be made over long periods of time in order to average tidal variations. However, no matter how long a period of data is averaged, the ideal true mean sea-level is unattainable because changes are taking place over short and long time scales".

To determine the mean sea surface without oceanographic and meteorological effects, in order to approach the "true" geoid level, hydrodynamic methods can be used (see Section 1.4.1), as for the measurement of steric changes. This method has shown that in northwest Europe there is a northward decrease of mean air pressure of approximately 0.4 mb per 100 km. However, this difference in air pressure is only equivalent to a 4 cm increase in adjusted mean sea level between Newlyn and the north of Scotland, the main effect being a raising of levels from west to east due to the winds associated with these mean atmospheric pressure gradients (Pugh 1987). Similar anomalies between the results of geodetic levelling and hydrodynamic levelling have been reported from other areas. Unfortunately, hydrodynamic observations made by oceanographers in relation to a sufficiently deep isobaric surface are limited by the occurrence of sills and straits, which create a discontinuity, and cannot extend to shallow coastal waters. This means that oceanographers cannot use differences in mean sea level measured along a coast to compute residual water movements, and land surveyors usually do not adjust their networks for mean sea-level anomalies. It has so far not been possible to reconcile the results of geodetic and hydrodynamic levelling in a way which satisfies the exponents of either discipline (Pugh 1987).

6.2.3.2   Satellite geodesy

During the last two decades, satellites have offered an unprecedented opportunity to obtain a global coverage of the ocean surface at regular time intervals. What is measured by satellites is the nadir H_1 distance from the satellite altimeter to the surface of the ocean at the instant of measurement (Fig. 114). To establish sea-level changes, what has to be determined is a change in H_2, whereas H_3 (geoid elevation) and H (elevation of the satellite above the reference ellipsoid) must be known accurately. H_2 includes all oceanographic effects, i.e. tide, currents and associated eddies, meteorological effects, steric effects and crustal movements. The accuracy

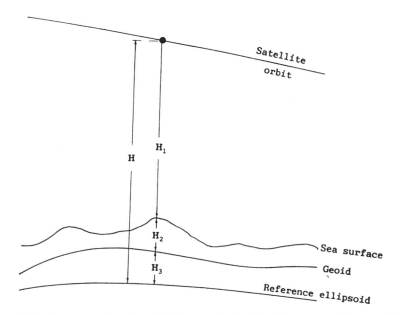

Figure 114   Parameters involved in interpreting height H_1 measured by a satellite
altimeter above the sea surface

of satellite altimetry is influenced by measurement errors and radial orbit errors. Ocean dynamic topography and tides also produce deviations from geoid form.

The first satellite provided with an altimeter, Skylab, launched in 1973, had a resolution of about 1 m. With the GEOS-3 satellite 2 years later, resolution improved to about 0.5 m, and with SEASAT in 1978 it reached 0.1 m (Pugh 1987). GEOSAT, launched in 1985, functioned successfully for almost 5 years of continuous operation, with a resolution similar to that of SEASAT, and demonstrated the potential of this method (Sandwell 1991).

The study of relative sea-level change in a given area is made possible only if a reference datum is chosen. Since most vertical datums are related to mean sea level, mean sea levels do not lie exactly on an equipotential surface owing to the effects of currents and other oceanographic phenomena, and local mean sea level diverges from the geoid by about ±1 m, the vertical datums of the world are only consistent at the ±1 m level.

In order to document present-day global sea-level changes, which may be of the order of millimetres per year, datums should be correlated at least on a decade basis at a ±1 cm level. Such a resolution is now being achieved with the ERS1 and TOPEX/POSEIDON satellite altimeters. The latter, for example, measured global mean sea level every 10 days over the first 2.5 years of the mission with a precision of 4 mm, and a rate of global sea-level rise of 5.8 ± 0.7 mm/yr was obtained. Such a rate is much faster than all the rates deduced from tide-gauge records (Table 4). The duration of observation is still too short, however, and a substantial portion of the trend observed may represent short-term variation attributable to El Niño episodes rather than to the effects of global warming (Nerem 1995a,b).

The required 1 cm vertical accuracy can also be obtained by continuous Global Positioning System (GPS) observations, to provide global tide-gauge station coordinates with day-to-day repeatability (Blewitt 1994). With this method, a consistent network of well-established tide gauges has been defined in the framework of the SELF (SEa Level Fluctuations) Project (Zerbini et al. 1996), encompassing the Mediterranean basin as far as the Black Sea. This has made it possible to determine to centimetre accuracy the tide-gauge benchmark heights in a global well-defined reference system such as the one provided by the Satellite Laser Ranging/Very Long Baseline Interferometry (SLR/VLBI) space techniques.

With the establishment of absolute datums with narrow uncertainty margins, one can expect major progress in the measurement of the current trends in sea-level change during the next decades.

6.3 EXPLANATION OF CURRENT ESTIMATIONS OF GLOBAL SEA-LEVEL RISE

To test the validity of the estimations summarized in Table 4 it is necessary to show how sea-level rises between 0.5 and 3.0 mm/yr over the last 100 years can be explained. According to the possible causes of sea-level change discussed in Chapter 1, steric changes and modifications in the quantity of oceanic water have to be considered.

Observational data related to thermal expansion are too sparse to make global-scale estimates, but model simulations have suggested a rise of the order of 0.3 ± 0.2 mm/yr due to steric changes (Table 6).

Although a few estimates of the mass budget of the Greenland and Antarctic ice sheets have been made (Table 7), continuous long-term measurements of the mass balances of glaciers and ice caps are very limited. According to Warrick and Oerlemans (1990 p. 268) "it is actually unknown how close the ice sheets are to equilibrium"; nevertheless, in case of a climatic warming, "one may expect an increasing surface mass balance for the Antarctic ice sheet (contributing to a sea-level lowering) and a decreasing mass balance for the other ice bodies (contributing to sea-level rise)". Accordingly, Antarctica is usually assumed to be in equilibrium or to have a slightly positive mass balance, the quantification of which is uncertain. The Greenland ice cap is probably close to balance (Meier 1984), with an uncertainty of ± 0.28 mm/yr of sea-level equivalent (Robin 1986). Many of the world's small glaciers have retreated over the last 100 years, but their contribution to sea-level rise is necessarily very limited (Table 2).

Table 6   Estimation of recent sea-level rise due to thermal expansion

Authors	Period of time considered	Sea-level rise (cm)	Rate of rise (mm/yr)
Etkins and Epstein (1982)	1890–1940	2.4	0.48
Gornitz et al. (1982); Gornitz and Lebedeff (1987)	1880–1980	1.6–2.8	0.14–0.45
Hoffman et al. (1983)	1900–1980	4.1	0.51
Barnett (1984)	1880–1980	<5	<5
Robin (1986)	1880–1980	2.3	0.23
Wigley and Raper (1987)	1880–1985	1.9–4.9	0.18–0.47
Warrick and Oerlemans (1990)	1890–1990	4±2	0.4±0.2

Table 7   Estimated effects on sea level of recent changes in ice volume (millimetres per year; + = sea-level rise)

Authors	Period of time considered	Antarctica	Greenland	Small glaciers
Meier (1984)	1900–1961			+0.46±0.26
US Dept Energy (1985)	unspecified	−0.6±0.6	−0.1±0.4	+0.5±0.3
Robin (1986)	1900–1975	−0.28	+0.3	+0.4
Klige and Dubrovolsky (1988)	1900–1975	+0.87	+0.23	+0.04
Warrick and Oerlemans (1990)	1890–1990	0.0±0.5	+0.25±0.15	+0.4±0.3

Altogether, it can be roughly estimated that glaciers and ice caps have accounted for 0.4 ± 0.3 mm/yr of the recent sea-level rise.

The effects of anthropic modification of land hydrology on sea level have been widely debated, with variable conclusions. For example, to mention two recent studies, Sahagian et al. (1994) found that the anthropic contribution to eustasy would be positive (a sea-level rise of 0.54 mm/yr), while Gornitz et al. (1996) thought it would be negative, with an upper limit corresponding (though with many uncertainties) to a sea-level fall of about 0.9 mm/yr.

Current estimates of changes in surface water and underground water storage are so uncertain and speculative, that, according to most recent IPCC conclusions, the most likely net contribution during the last 100 years has been close to zero, with an uncertainty of about ±0.5 mm/yr.

In summary, the most likely estimate is that all the above factors account for a global sea-level rise during the last 100 years of the order of 0.7 ± 1.0 mm/yr, i.e. near the lower end in the range of estimates in Table 4, though with a wide uncertainty band. Authors who claim that the recent global sea-level rise has been greater than 0.7 ± 1.0 mm/yr should explain the origin of the additional water volume necessary for such a rise.

6.4   COASTAL AREAS AT RISK
FROM SEA-LEVEL RISE

As sea level rises and water depth increases near the shore, wave strength increases, facilitating erosion. Erosion will therefore accelerate on coasts that were already retreating, and may be initiated on coasts that were previously stable or even prograding.

Analysis of the physical impacts of a sea-level rise is made more complex by the dense human occupation of many coastal areas, by their often high economic value and by the continuous interaction between human activities and natural processes. These complex aspects, including the potential effects of a rising sea level on coastal systems and human responses to a rising sea level, have been discussed in detail by Bird (1993).

The Caspian Sea coast, where sea level rose by 1.5 m between 1977 and 1990, may be considered a good natural laboratory in which to observe phenomena produced by a sea-level rise. Between 1977 and 1990, the extent of retreating beaches increased from 10% to 39% along the Caspian Sea coasts. Ignatov et al. (1993) observed various patterns of coastal zone development under sea-level rise (Fig. 115). On shallow shores, where wave energy dissipates on the gentle offshore slope, there is marine flooding of land without other dynamic changes in the coastal zone (Fig. 115a). Under higher inclinations, a long-shore barrier is developed near the zone where waves break, creating a lagoon (Fig. 115b). The most common pattern of shore evolution under inclinations of 0.005–0.01 is erosion of the upper offshore slope associated with wave reworking of deposits in this zone and the pushing of sediments towards the shoreline, where a barrier beach is formed, backed by a lagoon (Fig. 115c), or, if there is no lowland behind the beach ridge, a landward shift (Fig. 115d). Lastly, under slopes steeper than 0.01, the zone of erosion moves even higher landward and attacks the front of Holocene aggradation features (Fig. 115e).

Submergence of coastal areas at risk will not occur gradually, but by steps, after severe storms which open breaches in natural or artificial coastal defences. Climatic changes may also modify wind force and direction, altering the sedimentary budgets of longshore drift, increasing storm-surge frequencies and levels, and accelerating shoreline displacements.

Little is known about the local variations which would be induced by a global climatic change. Predictions of global climate models are often considered roughly acceptable on a broad scale, but reliable predictions of the time sequence of local and regional responses of variables such as temperature, wind and rainfall are not yet available. This usually prevents the making of credible predictions over a regional or local scale.

In moist tropical regions, an increase in air temperature would probably produce a migration of tropical-cyclone tracks towards areas where they are now less frequent. Impacts of more frequent tropical cyclones would be

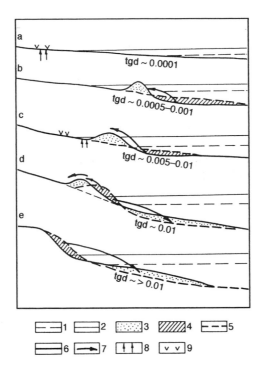

Figure 115   Model of transgressive evolution of the Caspian Sea coastal zone: 1, regressive sea level; 2, transgressive sea level; 3, accumulation of sediments; 4, erosion lens; 5, former profile of the coastal zone; 6, modern profile of the coastal zone; 7, transport of erosion material; 8, rise of ground water; 9, marshes (after Ignatov et al. 1993)

especially severe on coastal lowlands (tropical deltas and lagoons, coral islands).

In temperate coastal regions, however, storm surges are not expected to worsen greatly. Based on a scenario deduced from a general circulation model of the atmosphere, which predicts a 2° latitude northward displacement of extreme wind conditions in the North Sea, Bijl (1995) obtained only relatively minor changes (0.1 m) in the height of storm waves on the coasts of the Netherlands. On the other hand, an increase in the frequency and velocity of Sirocco winds has been observed during recent decades in the Adriatic Sea (Fig. 116). A further increase of this wind, which already causes frequent marine flooding in lowlands and around lagoons of the northwestern Adriatic, would be sufficient to endanger wide coastal areas, even in the absence of a sea-level rise. The evolution of coastal situations may therefore be quite variable, even in regions that are relatively close to each other.

Estimates and predictions have been proposed by various authors for the "global" sea-level rise, but there are many compelling arguments that on a

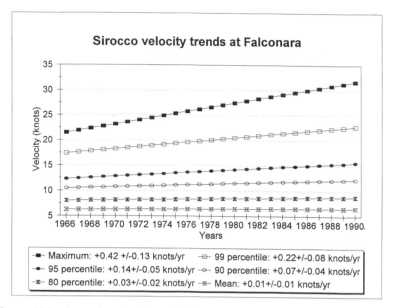

Figure 116 Annual velocities of Sirocco and southerly winds (120–220°) at Falconara, Ancona, Italy, from 1966 to 1990 (linear regressions). The increasing trends are statistically significant and stronger for the highest percentiles and especially for extreme events

regional basis the change in sea level is, and will be, different (e.g. Jeftic et al. 1992; Maul 1993). In addition, all these estimates are exclusive of vertical land movements and may have very little to do with actual changes in relative sea level experienced at a particular site. Many coastal areas are at present submerging (Fig. 117).

Figure 117 (*opposite*) Sectors of the world's coastline that have been subsiding in recent decades, as indicated by evidence of tectonic movements, increasing marine flooding, geomorphological and ecological indications, geodetic surveys, and groups of tide gauges recording a rise of mean sea level greater than 2 mm/yr over the past three decades (after Bird 1993). 1, Long Beach area; 2, Colorado River delta; 3, Gulf of La Plata; 4, Amazon delta; 5, Orinoco delta; 6, Gulf and Atlantic coasts; 7, southern and eastern England; 8, southern Baltic coasts; 9, southern coasts of the North Sea and Channel coasts; 10, Loire estuary; 11, Vendée coasts; 12, Lisbon region; 13, Gaudalquivir delta; 14, Ebro delta; 15, Rhône delta; 16, northern Adriatic low coasts of Italy; 17, Danube delta; 18, eastern Sea of Azov; 19, Poti Swamp; 20, southeast Turkey; 21, Nile delta to Libya; 22, northeast Tunisia; 23, Niger delta and north coasts of the Gulf of Guinea; 24, Zambezi delta; 25, Tigris–Euphrates delta; 26, Rann of Kutch; 27, southeastern India; 28, Ganges–Brahmaputra delta; 29, Irrawaddy delta; 30, Bangkok coastal region; 31, Mekong delta; 32, eastern Sumatra; 33, northern Java deltaic coast; 34, Sepik delta; 35, Port Adelaide region; 36, Corner Inlet region; 37, Hwang-He (Yellow River) delta; 38, head of Tokyo Bay; 39, Niigata; 40, Maizuru; 41, Manila; 42, Red River delta; 43, northern Taiwan

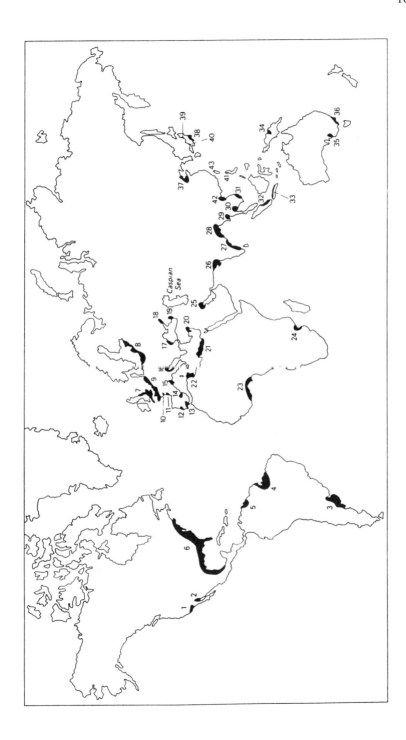

A global sea-level rise would indeed make most low-lying coastal areas more vulnerable to submergence and erosion. However, on actively submerging coasts, sea-level problems already exist, and these problems will probably worsen during the next century, even in the absence of an additional global sea-level rise.

In conclusion, the coastal areas most at risk in the near future, as sea level rises, will probably be those subsiding or below sea level, where problems already exist.

Conclusions

It was still believed in 1974, when the IGCP launched Project 61 "Sea-level changes during the last hemicycle", that following Suess' ideas it was possible to determine from field data a eustatic curve, valid for all areas. It is recognized today that the 1974 approach was unrealistic and misleading, because of the spatial sea-level variability, which depends on a combination of factors which operate globally, regionally or locally on different time scales.

In this book, various possible causes of sea-level change have been discussed. It has been stressed that climate changes have been the main cause of changes in quantity of oceanic water, especially during the last 20 ka, which correspond to the transition from a glacial to an interglacial stage.

Displacements of water masses are indeed a global phenomenon, in the sense that meltwater from former ice sheets tends to be distributed all through the oceans, to attain the same equipotential surface of the gravity field. This surface, far from being equidistant from the Earth's centre, as was believed for a long time, or even following closely the rotational ellipsoid which most approximates the Earth's surface, is made irregular and displaced continuously by many kinds of isostatic imbalances.

The imbalances resulting from glacio-isostasy cause vertical displacements of the Earth's surface which are often greater than glacio-eustatic effects in the near field of the ice sheets and ice-margin sites. This depends on the period considered, however, and the difference between glacio-isostatic uplift and glacio-eustatic sea-level rise may be positive during certain periods and negative during other periods, causing fluctuations in the relative sea level.

On intermediate and far-field sites, glacio-isostatic imbalances are much smaller than the eustatic factor, according to geophysical models, but are far from negligible. Hydro-isostatic imbalances tend to produce differential vertical movements on the coasts of the world, depending on local topography, width of the continental shelf, water depth and characteristics of the upper mantle.

Other kinds of isostatic imbalance (thermo-isostasy, volcano-isostasy, sediment-isostasy) may also be active, but usually have only local effects.

During the late Holocene, several thousand years after melting of the former ice sheets, glacio-isostatic uplift remains predominant at near-field and ice-margin sites. This situation will probably continue for several more thousand years, at exponentially decreasing rates, until an isostatic balance is established.

When the glacio-eustatic component of sea-level change greatly decreased, field data and global isostatic models show that relative sea-level histories on far-field sites differ from those on intermediate-field sites. At intermediate-field sites, where a peripheral bulge formerly surrounding the ice sheet tended to collapse, a slow relative sea-level rise is predominant, and subsidence is expected to continue in the future, though at exponentially decreasing rates. In far-field areas, a slight subsidence in the ocean basins (mainly of hydro-isostatic origin) and a comparable uplift of the continents (which are flexing upwards owing to the downward motion of the surrounding oceanic regions and the transfer of material from the region below the oceans to beneath the continents) have predominated during the last few thousand years. On continental far-field sites, remote from major active tectonic features, the maximum Holocene emergence often dates from about 5–6 ka BP and may reach 2–4 m (e.g., Figs. 79–81). At most intermediate-field sites, on the other hand, the present sea level is the highest level reached in the Holocene and is continuing to rise (e.g. Figs. 77, 78). In ocean regions in the far field of the ice sheets, hydro-isostatic subsidence may be compensated by a mechanism termed by Mitrovica and Peltier (1991) "equatorial ocean siphoning". This mechanism acts to draw water away from the equatorial regions, making some emergence possible, and is driven by the subsidence of those portions of the glacial forebulges which exist over oceanic regions. As subsidence proceeds, water is forced to flow into the peripheral regions (from the far field, and near-field regions undergoing uplift) in order that the oceans maintain hydrostatic equilibrium.

In a few thousand years there will be a difference in level between present-day shorelines in far-field and intermediate-field sites; accordingly, the elevation of the highest Holocene shoreline in far-field sites will be greater than today in relation to intermediate-field sites.

The differences in behaviour between far-field and intermediate-field sites have three major implications.

(1) Given the periodicities of astronomic origin which control global sea-level changes (Fig. 103), a complete isostatic balance will probably never be reached; when a new glaciation starts to make sea level fall again (perhaps within a few thousand years), relative sea level will still be rising on intermediate-field sites and falling on far-field sites; the relative vertical movements will simply change at intermediate-field sites from slight submergence to slight emergence soon after the beginning of the sea-level fall.

In far-field sites, on the contrary, the emergence rate will simply accelerate. In short, no coastal sector can remain vertically stable at any time (except for a short period when the maximum Holocene sea level is reached locally), and the present interglacial will leave features corresponding to its maximum sea level which may differ as much as a few metres in altitude and several thousand years in age from place to place.

(2) Situations similar to those of the present interglacial have probably occurred during past interglacials. It is therefore incorrect to ascribe great precision to sea-level elevations deduced from elevated marine terraces formed during preceding interglacial periods, and the development of marine terraces has not been perfectly synchronous everywhere.

(3) The fact that last-interglacial shorelines are found at about the same altitude at two sites in different areas does not necessarily mean that these two sites are "tectonically stable", or that they have experienced the same relative sea-level history during the late Quaternary; the rates of isostatic displacement may have been different at the two sites, although today they are at a similar vertical situation.

The term "eustasy", generally understood as "world-wide simultaneous change in sea level" (American Geological Institute 1960) is therefore an abstraction, referring to something which does not exist. Mörner (1986) noted that sea-level changes can no longer be claimed to be "worldwide" or "simultaneous" and proposed to redefine eustasy as "absolute sea level changes regardless of causation". Such redefinition has not been recommended by Fairbridge (1989), however, because it would embrace regional effects, destroying the essential point of the eustatic definition: "a universal rise or fall of sea level". It is in this wide sense, implying a change in the volume of oceanic waters, without implications of uniformity of vertical displacement or of simultaneity, that the terms "eustatic" or "glacio-eustatic" have been used in this book.

The most accurate sea-level indicators (discussed in Chapter 2), are certain *in situ* intertidal deposits, or upper sublittoral bioconstructions preserved in growth position. The narrowness of their vertical zonation and the possibility of dating by means of radiochronological techniques are essential elements to diminish uncertainty in the reconstruction of past relative sea-level histories. However, the temporal continuity of the reconstruction of a change in sea level depends also on the availability of such an indicator at different elevations.

Cores drilled in coral reefs seldom cover more than a few thousand years. The 52 m drill core reported by Chappell and Polach (1991) from the postglacial reef at Huon Peninsula, Papua New Guinea, showed that coral growth kept pace while relative sea level rose by 50 m, but the core only spans the interval from 7 to 11 ka BP. No section has yet been reported where the same sea-level indicator permits the local sea-level history to be determined from the last glacial maximum to the present. The longest continuous record is probably that reported from the barrier reef of Tahiti (Bard et al. 1996), where continuous coral reef growth at depths less than 6 m occurred during the last 13.8 ka, starting from the basaltic substratum at about 87 m depth. The most complete record based on a single sea-level indicator (in Barbados; curves E and F in Fig. 66) was obtained from compilation of three separate cores, and each segment of sea-level curve deduced from one of these cores is offset from the next.

More precise sea-level indicators such as the vermetid *Dendropoma petraeum* or the calcareous algae *Neogoniolithon notarisii* or *Lithophyllum lichenoides*, which may be used in the Mediterranean to reconstruct former sea levels with an accuracy of ±0.2 m or even ±0.1 m, have been observed in the same section covering time spans of no more than the last 4.5 ka on the coasts of southern France (Laborel et al. 1994) or the last 4.2 ka in Crete (Pirazzoli et al. 1982). An accuracy of the same order (±0.1 m) has been reported from macrotidal palaeomarsh deposits using certain foraminiferal assemblages (Scott and Medioli 1986), but for a marsh with such precise assemblages to be maintained, a delicate balance is necessary between the rate of relative sea-level rise and that of sediment accretion, which seldom remains unchanged during long periods.

In most other cases, relative sea-level curves have to be based on discontinuous evidence (e.g. on peat layers or on other datable material from drilled cores) or on a variety of more or less precise sea-level indicators sampled from different sections, or even on discontinuous archaeological or historical information.

Reliable data on the sea-level position during the last glacial maximum are very limited. Such information is totally absent from formerly glaciated areas. In Fennoscandia, the uplift since the last glacial peak is estimated to have reached a maximum of 800–850 m (Mörner 1979), i.e. several times more than the eustatic sea-level rise. Even greater uplift probably occurred in northern Canada.

Evidence for the sea-level position during the last glacial maximum can be obtained only from underwater intermediate-field or far-field sites. Some data (summarized in Chapter 3) suggest that the shoreline of the glacial maximum was situated at depths of between 110 and 175 m, the differences being ascribed to non-uniform isostatic responses to the superimposed effects of deglaciation unloading and meltwater load.

A major consequence of the low eustatic sea level during the last glacial maximum is that many continental-shelf areas became exposed land, creating land bridges between continents that are now separated by the sea. This was the case, in particular, of the Bering Strait (Fig. 53 (Plate I) and Fig. 54a (Plate II)), which enabled human and faunal migrations between Asia and North America. The development of continental ice sheets and of mountain glacial complexes made most of the North American continent unsuitable for human occupation during the glacial maximum, thus making migrations possible only through ice-free corridors which must have existed during certain milder climate interstadials.

Other areas of major land bridges during the last glacial maximum were located on the Sunda and Sahul shelves (Fig. 54c (Plate II)), controlling human migrations between southeast Asia and New Guinea–Australia, in the North Sea area (Fig. 54d (Plate II)) and in various areas of the Mediterranean (Fig. 57).

The global sea-level rise accompanying the last deglaciation started some 17 radiocarbon ka BP, i.e. about 20 ka ago. The most detailed information is available from Barbados for the whole deglacial period (Fairbanks 1989; Bard et al. 1990b; see Fig. 66, curves E and F), from Papua New Guinea between 7 and 11 ka BP (Chappell and Polach 1991), i.e. between 8 and 13 ka ago (Edwards et al. 1993), and from Tahiti for the last 13.8 ka (Bard et al. 1996). Such information shows, in spite of small but systematic differences which can be attributed to different glacio- and hydro-isostatic responses, similar trends, which probably have a global significance.

Outside near-field and ice-margin sites, the last deglaciation appears to have been characterized by a smooth and continuous eustatic rise of sea level with no reversals, including two brief periods of accelerated melting. These periods of accelerated sea-level rise have been dated at about 14.0 ka and 11.3 ka ago, respectively, and are thought to be related to massive inputs of continental ice, corresponding to 50–40 and 30–20 mm/yr of global sea-level change (Bard et al. 1996). These two sea-level leaps, especially the first one, were rapid enough to drown coral reefs and other coastal features in most places, often creating hiatuses and unconformities in the fossil coastal record, even in rapidly uplifting areas such as the east coast of Taiwan (Fig. 118).

A detailed image for the climate change in the North Atlantic accompanying the deglacial history can be obtained by plotting on the same calibrated (absolute) time scale, the $\delta^{18}O$ of the GRIP Greenland ice core (Johnsen et al. 1992) and the $\delta^{18}O$ record of deep-sea core SU81-18 collected at about 38°N in the northeast Atlantic (Bard et al. 1987). The curves shown in Fig. 119 suggest that it was indeed the warm climate of the early Bölling period which accelerated the ice-sheet melting, producing a melting surge at 13.8–14.1 ka ago; in the same way, it was the warm climate of the early Preboreal, after the cool Younger Dryas episode, which caused a second meltwater pulse about 11 ka ago. During the two colder periods, aridity increased in the Sahara and Sahel regions of North Africa (Gasse et al. 1990).

Although in intermediate-field and far-field sites deglacial times were characterized by a sea-level rise without reversals, this was not usually the case at near-field and ice-margin sites. Near former ice sheets the local trend in relative sea-level change depended at any time on the resultant of two major components: the glacio-eustatic sea-level rise, and the glacio-

Figure 118   A layer of *in situ* fossiliferous beach or shallow marine deposits dated from about 14 ka ago has been uplifted at 17 m in altitude (at the level of the observer's eye) near Tulan, east coast of Taiwan, by an uplift rate estimated at 7.6 ± 0.9 mm a^{-1}. These deposits are capped by an unconformity and by relatively recent river-mouth deposits with cobbles. The rapid increase in the water depth a short time after deposition has probably favoured the preservation of the coastal deposits below the unconformity (from Pirazzoli et al. 1993; Photo B675, Jan. 1990)

isostatic land uplift. The first component was not a uniform phenomenon, but had episodes of acceleration and deceleration, as discussed above, or even of relative stability, for example during the last 7 ka. The second component followed roughly an exponentially decreasing trend, at rates varying with the distance from the ice sheet barycentre and depending on the ice-sheet mass and shape. The superimposition of these two major components resulted in local fluctuations in the relative sea level near the ice margins, at amplitudes changing from place to place, some of which have been documented during the Younger Dryas episode.

The relative sea-level curves compiled in Figs. 65–69 summarize available information on local sea-level changes at selected sites or areas during the last 20 ka. At first sight, these curves show great diversity, which may be explained in part by different isostatic and tectonic conditions, but also by differences in methodological approach, by the use of more or less reliable sea-level indicators, by inadequate vertical resolution or even by the absence of estimation of uncertainty margins, and, in some cases, by certain arbitrary interpretations. These curves are discussed in more detail in Chapter 4. The main problem is that, in spite of recognized and sustained

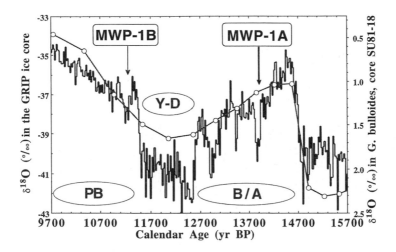

Figure 119 Oxygen isotope proxy data in the North Atlantic between 9.7 and 15.7 ka ago (absolute ages). The grey solid line represents the $\delta^{18}O$ variations measured for the GRIP ice core and the open dots show the planktonic foraminifera $\delta^{18}O$ record of deep sea core SU81-18. A strong warming ($\delta^{18}O$ increase at about 14.5 ka ago in the Greenland ice) at the beginning of the Bölling-Alleröd (B/A) period is synchronous with the sharp warming in core SU81-18 ($\delta^{18}O$ decrease in the planktonic foraminifera). A similar agreement can be seen for the Younger Dryas (Y-D) cold period and Preboreal (PB) warming. MWP-1A and MWP-1B: major meltwater pulses (adapted from Bard et al. 1996)

interest, reconstructing the variations of sea level for the last 20 ka has proved to be a difficult task. Much work remains to be done before relative sea-level changes can be determined in detail for all geographical areas, especially for the period from 20 to 10 ka BP.

When many continental shelves emerged during the last glacial maximum, the total area of terrestrial surfaces were increased by over 10%. More information on the shoreline positions at different periods, enabling definition of the continental-shelf surfaces exposed at different stages of the transgression, and on the physical, ecological and geochemical impacts that accompanied the submergence of these areas, would certainly be useful to the development of global change studies.

With the decrease of eustatic and isostatic components, other minor causes of sea-level variations may appear. Vertical land displacements and tilting of shorelines of tectonic origin may become a predominant factor of relative sea-level change near tectonic plate boundaries. Coseismic vertical movements may leave sequences of stepped Holocene shorelines (Fig. 120; see also Section 5.3), or belts of marine bioconstructions well-preserved above the midlittoral zone (Fig. 121). Storm surges and tsunamis may also leave recognizable features in the sedimentary coastal record.

Figure 120   Nine erosional shorelines appear between 1.1 and 2.7 m above sea level
on a limestone cliff in Antikythira Island (Greece). They correspond to a series of
coseismic small sinkings, at intervals of about 200–250 years, which took place
between about 4 and 1.7 ka BP. A final coseismic uplift of 2.7 m, raising at one time
the whole sequence of shorelines, is ascribed to a great earthquake which occurred in
AD 365. A metre rule gives scale (Photo 5029, Sep. 1979)

In short, during the relative eustatic stability which has characterized the
last 6–7 ka, many minor causes of relative sea-level change have become
significant, forming a "sea-level noise" which tends to mask the purely
eustatic factor.

A similar situation is found when recent changes in sea level are considered. Trends indicated by tide-gauge records are extremely variable from
place to place, indicating at each time the predominance of a "noise" (on
the order of centimetres per year) of isostatic, tectonic, hydrodynamic,
climatic or human-induced origin, whereas the eustatic factor (on the order
of millimetres per year) remains undetermined. Although a great number of
tide-gauge stations are in use, most records are not long enough to present
statistically significant secular trends, and those which are long enough are

Figure 121   Stepping rims of the vermetid *Dendropoma* and the calcareous alga *Neogoniolithon* (west coast of Crete, Greece) enabled a series of small coseismic movements (the same as that which displaced the erosional features shown in Fig. 120) to be dated by radiocarbon (Photo 4760, May 1978)

too unevenly distributed on the Earth's surface to make possible a precise estimation of the global trend in present-day sea-level change. It is not surprising, therefore, that attempts to estimate such global trends from tide-gauge records have given a wide range of results, from 0.5 to 3.0 mm/yr (Table 4).

Broad variations on the ocean surface may depend on an uneven mass distribution inside the Earth or on load distribution on the Earth's surface, which are relatively stable over time scales of 100 years or less. On the contrary, minor slopes show a great variability at periods of several years or decades, which is due to wind stress and ocean current variability modulating ocean topography (Bilham 1991). This is true not only near major oceanic currents (Figs. 9, 111, 112), but also on a more global scale. For example, Gröger and Plag (1993) used available tide-gauge records covering the period 1951–1989 to calculate the trend differences for estimates of a 19 year moving window (to filter the influence of the 18.61 year lunar declination cycle and that of the Saros cycle), shifted by 9.5 years (Fig. 122). The differences between the trends of the interval 1951–1970 and those of 1969–1979 (Fig. 122a) indicate an acceleration for most of the eastern coasts of America and Asia (northwestern Atlantic and Pacific Oceans, respectively), while the trends of the western coasts of

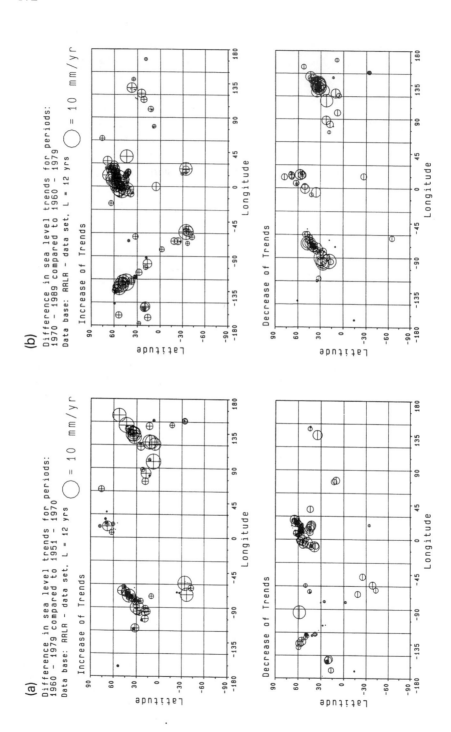

(a) Difference in sea level trends for periods:
1960 – 1979 compared to 1951 – 1970
Data base: RLR – data set. L = 12 yrs ◯ = 10 mm/yr

(b) Difference in sea level trends for periods:
1970 – 1989 compared to 1960 – 1979
Data base: RLR – data set. L = 12 yrs ◯ = 10 mm/yr

America and Eurasia (eastern sides of the Atlantic and Pacific Oceans) have generally been decreasing, with the decreases most pronounced at the European coasts. This picture is nearly inverted for the differences between 1960–1979 and 1970–1989 (Fig. 122b). Most of the stations on western coasts have accelerated trends while the trends at stations located on eastern coasts have been reduced. Such a spatial and temporal pattern is in agreement with a slow global east–west motion of the ocean heights on decadal to interdecadal time scales, resulting in a total change of mean sea level of up to 10 cm within the period from 1951 to 1989.

Such variability of slopes on the ocean surface with time should be taken into account when estimates of rates of recent global sea-level change are attempted, but this was not the case for most estimates summarized in Table 4.

It has been shown in Section 6.3 that only a limited range of global sea-level changes can be explained by steric changes and by modifications in the quantity of oceanic water, and that recent global sea-level changes greater than 0.7 ± 1.0 mm/yr are unlikely. In the opinion of the author, the most realistic range of recent global sea-level rise may even be restricted to 0.9 ± 0.3 mm/yr. It is interesting to note that nobody has claimed that the global sea level has been falling during the last 100 years, and no evidence can yet be found for mean sea-level rise accelerations.

It may be deduced from several parts of this book that in many domains the knowledge available is still scanty, in spite of important improvements which have been made during the last decades by accurate field work, more precise dating methods, and new geophysical modelling. If the spatial variability of relative sea-level changes has been demonstrated, and the main causes of this variability roughly identified, detailed field verifications have been carried out only in a few places. Of the last 20 ka, the period from 20 to 10 ka BP is indeed the least well known. The mapping of world shorelines during the last glacial maximum could be a stimulating task for future research. This work should be based not only on modelling un-corrected for tectonic displacements (Edwards 1995) (e.g. Figs. 53, 54, 56), or on the assumption (which also deforms the reality) that the shoreline was situated along a uniform bathymetric curve, but verified by field data and, where possible, accurately dated. Pioneering work in this direction has

---

Figure 122 (*opposite*) Temporal changes of trend pattern using stations with records of at least 12 years. The size of the circles is proportional to the amount of the difference. The sign (+/–) is indicated inside the circles. A positive difference indicates an acceleration trend. (a) Differences for the intervals 1951–1970 and 1960–1979. (b) Differences for the intervals 1960–1979 and 1970–1989. For the sake of clarity, positive and negative differences have been plotted separately in the upper and lower part, respectively. Note the distribution of the signs, which in (a) is mainly positive for the eastern and negative for the western coasts, with a nearly inverted pattern in (b) (after Gröger and Plag 1993)

already been undertaken by Russian oceanographers in areas such as the Black Sea (Fig. 58), the Seychelles Bank (Fig. 60), and North Eurasia (Biryukov et al. 1988), and by Dutch geologists in the North Sea (Fig. 97). Similar work should be extended to other regions and could form the theme of new international cooperation projects, such as the IGCP Project 396, which was approved by UNESCO and IUGS in 1996.

The identification of the 18 ka BP shoreline could be the first step of wider research aimed at recognizing in more detail the palaeogeography of continental shelves and the various natural resources which may have been concentrated by the deglacial sea-level rise. Former shorelines corresponding to characteristic periods preceding and following rapid episodes of eustatic sea-level rise (e.g. before and after the major meltwater pulses of about 14 and 11 ka ago) should also be mapped in detail, because of their major palaeogeographic and palaeoecological implications. The existence of such maps would be useful to deduce the time and direction of past human and mammal migrations, to estimate the contribution of continental shelves to global carbon budgets between a glacial and an interglacial period, to make possible a better exploitation of natural resources submerged on the continental shelves, and to localize underwater sites of prehistoric value.

Another field in which improvements would be most useful is in distinguishing vertical land movements from eustatic movements, especially for recent times, in order to improve our ability to estimate the present-day rate of global sea-level rise and its evolution in time. The absence of vertically stable reference datums has until now prevented accurate estimates of global sea-level change, at practically all time scales (Pirazzoli 1993). Fortunately, such improvements are already under way, thanks to the latest generation of geodetic satellites, and major progress can be expected during the next decades. New data from satellite altimeters are expected to contribute substantially to our understanding of the distribution of ocean currents and their effect on coastal sea slopes (Bilham 1991) and to the establishment of present-day rates of absolute sea-level change.

References

Adams, J.M., Faure, H., Faure-Denard, L., McGlade, J.M. and Woodward, F.I. (1990) Increases in terrestrial carbon storage from the Last Glacial Maximum to the present. *Nature*, **348**, 711–714.

Adey, W.H. (1986) Coralline algae as indicators of sea level. In: O. Van de Plassche (ed.) *Sea-Level Research: a Manual for the Collection and Evaluation of Data*. Geo Books, Norwich, pp. 229–280.

Alessio, M., Allegri, L., Antonioli, F., Belluomini, B., Ferranti, L., Improta, S., Manfra, L. and Proposito, A. (1992) Risultati preliminari relativi alla datazione di speleotemi sommersi nelle fasce costiere del Tirreno centrale. *Giornale di Geologia*, **54**(2), 165–193.

Alessio, M., Allegri, L., Antonioli, F., Belluomini, B., Improta, S., Manfra, L. and Martinez, M.P. (1994) La curva di risalita del mare Tirreno negli ultimi 40 ka ottenuta mediante datazioni di speleotemi sommersi e dati archeologici. In: F. Antonioli, C. Donadio and L. Ferranti (eds.) *International Meeting on Underwater Geology (June 8–10, 1994, Palinuro) – Guide to the Excursion, Abstracts*. Universitá Napoli and ENEA, Naples, pp. 74–75.

Alley, R.B. et al. (11 authors) (1993) Abrupt increase in Greenland snow accumulation at the end of the Younger Dryas event. *Nature*, **362**, 527–529.

Allighieri, D. (1320) *De aqua et terra. Quistione trattata in Verona da Dante Allighieri il Di 20, Gennajo MCCCXX intorno alla forma del globo terracqueo ed al luogo rispettivamente occupato dall'acqua e dalla terra*. Torri, Livorno, 1843.

Aloïsi, J.C., Monaco, A., Planchais, N., Thommeret, J. and Thommeret, Y. (1978) The Holocene transgression in the Golfe du Lion, southwestern France: paleogeographic and paleobotanical evolution. *Géographie Physique et Quaternaire*, **32**(2), 145–162.

American Geological Institute (1960) *Glossary of Geology and Related Sciences*, 2nd edn. Washington, DC.

André, M.F. (1993) *Les Versants du Spitsberg*. Presses Université Nancy, Nancy, 361 pp.

Andrews, J.T. (1986) Elevation and age relationships: raised marine deposits and landforms in glaciated areas. In: O. Van de Plassche (ed.) *Sea-Level Research: a Manual for the Collection and Evaluation of Data*. Geo Books, Norwich, pp. 67–95.

Anundsen, K. (1985) Changes in shore-level and ice-front position in Late Weichsel and Holocene, southern Norway. *Norsk Geografisk Tidsskrift*, **39**, 205–225.

Atwater, B.F. (1987) Evidence for great Holocene earthquakes along the outer coast of Washington State. *Science*, **236**, 942–944.

Atwater, B.F. and Moore, A.L. (1992) A tsunami about 1000 years ago in Puget Sound, Washington. *Science*, **258**, 1614–1617.

Bada, J.L. and Finkel, R. (1983) The upper Pleistocene peopling of the new world: evidence derived from radiocarbon, amino acid racemization and uranium series dating. In: P.M. Masters and N.C. Flemming (eds.) *Quaternary Coastlines and Marine Archaeology*. Academic Press, London, pp. 463–479.

Badyukov, D.D., Demidenko, E.L. and Kaplin, P.A. (1989) Paleogeography of the Seychelles Bank and the northwest Madagascar shelf during the last glacio-eustatic regression (18,000 B.P.). *Chinese Journal of Oceanology and Limnology*, **7** (1), 89–92.

Baeteman, C. (1985) Late Holocene geology of the Marathon plain (Greece). *Journal of Coastal Research*, **1**, 173–185.

Bailey, G. and Parkington, J. (eds.) (1988) *The Archaeology of Prehistoric Coastlines*. Cambridge University Press, Cambridge, 154 pp.

Baker, T.F. et al. (19 authors) (1995) Sea-level fluctuations: geophysical interpretation and environmental impact (SELF). In: I. Troen (ed.) *Climate Change and Impacts, Proceedings of the Symposium held in Copenhagen, Denmark, Sept. 6–10, 1993*. European Commission, Science Research Development, EUR 15921 EN, pp. 323–338.

Bard, E., Arnold, M., Maurice, P., Duprat, J., Moyes, J. and Duplessy, J.C. (1987) Retreat velocity of the North Atlantic polar front during the last deglaciation determined by ^{14}C accelerator mass spectrometry. *Nature*, **328**, 791–794.

Bard, E., Hamelin, B. and Fairbanks, R.G. (1990a) U-Th ages obtained by mass spectrometry in corals from Barbados: sea level during the past 130,000 years. *Nature*, **346**, 456–458.

Bard, E., Hamelin, B., Fairbanks, R.G. and Zindler, A. (1990b) Calibration of the ^{14}C timescale over the past 30 000 years using mass spectrometric U-Th ages from Barbados corals. *Nature*, **345**, 405–410.

Bard, E., Arnold, M., Fairbanks, R.G. and Hamelin, B. (1993) ^{230}Th-^{234}U and ^{14}C ages obtained by mass spectrometry on corals. *Radiocarbon*, **35** (1), 191–199.

Bard, E., Hamelin, B., Arnold, M., Montaggioni, L., Cabioch, G., Faure, G. and Rougerie, F. (1996) Deglacial sea level record from Tahiti corals and the timing of global meltwater discharge. *Nature*, **382**, 241–244.

Barham A.J. and Harris, D.R. (1983) Prehistory and palaeoecology of Torres Strait. In: P.M. Masters and N.C. Flemming (eds.) *Quaternary Coastlines and Marine Archaeology*. Academic Press, London, pp. 529–557.

Barnett, T.P. (1983) Recent changes in sea level and their possible causes. *Climatic Change*, **5**, 15–38.

Barnett, T.P. (1984) The estimation of "global" sea level change: a problem of uniqueness. *Journal of Geophysical Research*, **89** (C5), 7980–7988.

Barusseau, J.P., Descamps, C., Giresse, P., Monteillet, J. and Pazdur, M. (1989) Nouvelle définition des niveaux marins le long de la côte nord-mauritanienne (Sud du Banc d'Arguin) pendant les cinq derniers millénaires. *Comptes Rendus de l'Académie des Sciences, Paris*, **309** (II), 1019–1024.

Beets, D.J., Rijsdijk, K.F., Laban, C. and Cleveringa, P. (in press) Holocene sea-level rise and shoreline positions in the southern North Sea. *Marine Geology*.

Behre, K.E. (1986) Analysis of botanical macro-remains. In: O. Van de Plassche (ed.) *Sea-Level Research: a Manual for the Collection and Evaluation of Data*. Geo Books, Norwich, pp. 413–431.

Berger, A. (1988) Milankovitch theory and climate. *Reviews of Geophysics*, **26** (4), 624–657.

Berger, A. and Loutre, M.F. (1996) Modelling the climate response to astronomical and CO_2 forcings. *Comptes Rendus de l'Académie des Sciences Paris*, **322** (IIa) (in press).

Berger, A., Gallée, H., Fichefet, T., Marsiat, I. and Tricot, C. (1990) Testing the astronomical theory with a coupled climate–ice-sheet model. *Global and Planetary Change*, **3**, 125–141.

Bernier, P., Bonvallot, J., Dalongeville, R. and Prieur, A. (1990) Le beach-rock de Temae (Ile de Moorea – Polynésie française) – Signification géomorphologique et processus diagénétiques. *Zeitschrift für Geomorphologie*, **34**, 435–450.

Berrino, G., Corrado, G., Luongo, G. and Toro, B. (1984) Ground deformation and gravity changes accompanying the 1982 Pozzuoli uplift. *Bulletin Volcanologique*, **47** (2), 187–200.

Berryman, K.R., Ota, Y. and Hull, A.G. (1989) Holocene paleoseismicity in the fold and thrust belt of the Hikurangi subduction zone, eastern North Island, New Zealand. *Tectonophysics*, **163**, 185–195.

Bijl, W. (1995) *Impact of a Wind Climate Change on the Surge in the Southern Part of the North Sea*. Report RIKZ-95.016, National Institute for Coastal and Marine Management, Den Haag, 43 pp.

Bilham, R. (1991) Earthquakes and sea level: space and terrestrial metrology on a changing planet. *Reviews of Geophysics*, **29**, 1–29.

Bindoff, N.L. and Church, J.A. (1992) Warming of the water column in the southwest Pacific Ocean. *Nature*, **357**, 59–62.

Bird, E.C.F. (1988) The tubeform *Galeolaria caespitosa* as an indicator of sea level rise. *Victorian Naturalist*, **105**, 98–104.

Bird, E.C.F. (1993) *Submerging Coasts – The Effects of a Rising Sea Level on Coastal Environments*. Wiley, Chichester, 184 pp.

Bird, E.C.F. (1996) *Beach Management*. Wiley, Chichester, 281 pp.

Bird, E.C.F. and Klemsdal, T. (1986) Shore displacement and the origin of the lagoon at Brusand, southwestern Norway. *Norsk Geografisk Tidsskrift*, **40**, 27–35.

Biryukov, V.Y., Faustova, M.A., Kaplin, P.A., Pavlidis, Y.A., Romanova, E.A. and Velichko, A.A. (1988) The paleogeography of Arctic shelf and coastal zone of Eurasia at the time of the last glaciation (18,000 yr B.P.). *Palaeogeography, Palaeoclimatology, Palaeoecology*, **68**, 117–125.

Blackwelder, B.W., Pilkey, O.H. and Howard, J.D. (1979) Late Wisconsinan sea levels on the southeast U.S. Atlantic shelf based on in-place shoreline indicators. *Science*, **204**, 618–620.

Blewitt, G. (1994) The Global Positioning System. In: W.E. Carter (ed.) *Fixing Committee, Deacon Laboratory, Godalming, Surrey, U.K., December 13–15, 1993*. NOAA Technical Report NOSOES0006, pp. 17–26.

Bloom, A.L. (1967) Pleistocene shorelines: a new test of isostasy. *Geological Society of America Bulletin*, **78**, 1477–1494.

Bloom, A.L. (1971) Glacial-eustatic and isostatic controls of sea level since the last glaciation. In: K.K. Turekian (ed.) *The Late Cenozoic Glacial Ages*. Yale University Press, New Haven and London, pp. 355–379.

Bloom, A.L. and Yonekura, N. (1990) Graphic analysis of dislocated Quaternary shorelines. In: *Sea-Level Change*. Studies in Geophysics. National Academy Press, Washington DC, pp. 104–115.

Bondesan, M., Castiglioni, G.B., Elmi, C., Gabbianelli, G., Marocco, R., Pirazzoli, P.A. and Tomasin, A. (1995) Coastal areas at risk from storm surges and sea-level rise in northeastern Italy. *Journal of Coastal Research*, **11** (4), 1354–1379.

Broecker, W.S. (1994) Massive iceberg discharges as triggers for global climate changes. *Nature*, **372**, 421–424.

Broecker, W.S., Peteet, D.M. and Rind, D (1985) Does the ocean-atmosphere system have more than one stable mode of operation? *Nature*, **315**, 21–26.

Broecker, W.S., Kennett, J.P., Flower, B.P., Teller, J.T., Trumbore, S., Bonani, G. and Wolfli, W. (1989) Routing of meltwater from the Laurentide ice sheet during the Younger Dryas cold episode. *Nature*, **341**, 318–321.

Brookes, I.A., Scott, D.B. and McAndrews, J.H. (1985) Postglacial relative sea-level change, Port au Port area, west Newfoundland. *Canadian Journal of Earth Science*, **22**, 1039–1047.

Bryan, E.H. (1953) Check list of atolls. *Atoll Research Bulletin*, **19**, 1–38.

Bucknam, R.C., Hemphill-Haley, E. and Leopold, E.B. (1992) Abrupt uplift within the past 1700 years at southern Puget Sound, Washington. *Science*, **258**, 1611–1614.

Cabioch, G., Montaggioni, L.F. and Faure, G. (1995) Holocene initiation and development of New Caledonian fringing reefs, SW Pacific. *Coral Reefs*, **14**, 131–140.

Cailleux, A. (1952) Récentes variations du niveau des mers et des terres. *Bulletin de la Société Géologique de France*, série 6, **2**, 135–144.

Cailleux, A. (1969) *Les Sciences de la Terre*. Bordas, Paris, 799 pp.

Campbell, J.F. (1986) Subsidence rates from the southeastern Hawaiian islands determined from submerged terraces. *Geo-Marine Letters*, **6**, 139–146.

Caputo, M., Pieri, L. and Unguendoli, M. (1970) Geometric investigation of the subsidence in the Po Delta. *Bollettino di Geofisica Teorica ed Applicata*, **13** (47), 187–207.

Cathles, L.M. III (1975) *The Viscosity of the Earth's Mantle*. Princeton University Press, Princeton, NJ, 386 pp.

Chappell, J. and Polach, H. (1991) Post-glacial sea-level rise from a coral record at Huon Peninsula, Papua New Guinea. *Nature*, **349**, 147–149.

Chappell, J. and Shackleton, N.J. (1986) Oxygen isotopes and sea level. *Nature*, **324**, 137–140.

Chappell, J., Chivas, A., Wallensky, E., Polach, H.A. and Aharon, P. (1983) Holocene palaeo-environmental changes, central to north Great Barrier Reef inner zone. *BMR (Bureau of Mineral Resources) Journal of the Australian Geology and Geophysics*, **8**, 223–235.

Chappell J., Omura, A., McCulloch, M., Esat, T., Ota, Y. and Pandolfi, J. (1994) Revised Late Quaternary sea levels between 70 and 30 ka from coral terraces at Huon Peninsula. In: Y. Ota (ed.) *Study on Coral Reef Terraces of the Huon Peninsula, Papua New Guinea*. Department of Geography, Yokohama National University, pp. 155–166.

Cheney, R.E., Marsh, J.G. and Beckley, B.D. (1983) Global mesoscale variability from collinear tracks of SEASAT altimeter data. *Journal of Geophysical Research*, **88** (C7), 4343–4354.

Chepalyga, A.L. (1984) Inland sea basins. In: A.A. Velichleo, H.E. Wright and C.W. Barnosky (eds.) *Late Quaternary Environments of the Soviet Union*. Longman, London, pp. 229–247.

Clark, J.A. and Lingle, C.S. 1979. Predicted relative sea-level changes (18,000 years B.P. to Present) caused by late-glacial retreat of the Antarctic ice sheet. *Quaternary Research*, **11**, 279–298.

Clark, J.A., Farrell, W.E. and Peltier, W.R. (1978) Global changes in postglacial sea level: a numerical calculation. *Quaternary Research*, **9**, 265–278.

Clarke, S.H. and Carver, G.A. (1992) Late Holocene tectonics and paleoseismicity, southern Cascadia subduction zone. *Science*, **255**, 188–192.

CLIMAP Project Members (1976) The surface of the ice-age earth. *Science*, **191**, 1131–1137.

COHMAP Members (1988) Climatic changes of the last 18,000 years: observations and model simulations. *Science*, **241**, 1043–1052.

Colantoni, P., Gallignani, P. and Lenaz, R. (1979) Late Pleistocene and Holocene evolution of the north Adriatic continental shelf (Italy). *Marine Geology*, **33**, M41–M50.

Colantoni, P., Preti, M. and Villani, B. (1990) Sistema deposizionale e linea di riva olocenica sommersi in Adriatico al largo di Ravenna. *Giornale di Geologia*, **52** (1), 1–18.

Collina-Girard, J. (1995) La grotte Cosquer et les sites paléolithiques du littoral marseillais (entre Carry-le Rouet et Cassis). *Méditerranée*, **82** (3–4), 7–19.

Colonna, M. (1994) *Chronologie des variations du niveau marin au cours du dernier cycle climatique (0–140000 ans) dans la partie sud occidentale de l'Océan Indien*. Thèse de Doctorat, Université de Provence (Aix-Marseille I), 293 pp.

Conchon, O. et al. (15 authors) (1992) Glaciation and permafrost. In: B. Frenzel, M. Pécsi and A.A. Velichko (eds.) *Atlas of Paleoclimates and Paleoenvironments of the Northern Hemisphere*. Geographical Research Institute, Hungarian Academy of Sciences, Budapest, plate 49.

Coudray, J. and Delibrias, G. (1972) Variations du niveau marin au-dessus de l'actuel en Nouvelle Calédonie depuis 6000 ans. *Comptes Rendus de l'Académie des Sciences, Paris*, **275**, 2623–2626.

Coudray, J. and Montaggioni, L. (1986) The diagenetic products of marine carbonates as sea-level indicators. In: O. Van de Plassche (ed.) *Sea-Level Research: A Manual for the Collection and Evaluation of Data*. Geo Books, Norwich, pp. 311–360.

Cuffey, K.M., Clow, G.D., Alley, R.B., Stuiver, M., Waddington, E.D. and Saltus, R.W. (1995) Large Arctic temperature change at the Wisconsin–Holocene glacial transition. *Science*, **270**, 455–458.

Curray, J.R. (1961) Late Quaternary sea level: a discussion. *Geological Society of America Bulletin*, **72**, 1707–1712.

Curray, J.R. (1965) Late Quaternary history, continental shelves of the United States. In: Whuglit and Frey (eds.) *The Quaternary of the United States*. VII INQUA Congress, Princeton University Press, Princeton, NJ, pp. 723–735.

Dabrio, C.J, Zazo, C., Goy, J.L., Santisteban, C. de, Bardají, T. and Somoza, L. (1991) *Neogene and Quaternary Fan-delta Deposits in Southeastern Spain – Field Guide*. Cuadernos de Geología Ibérica, Madrid, No. 15, pp. 327–400.

Dalongeville, R. (ed.) (1984) *Le Beach-Rock*. Travaux de la Maison de l'Orient, Lyon, vol. 8, 197 pp.

Dalongeville, R., Laborel, J., Pirazzoli, P.A., Sanlaville, P., Arnold, M., Bernier, P., Evin, J. and Montaggioni, L.F. (1993) Les variations récentes de la ligne de rivage sur le littoral syrien. *Quaternaire*, **4**, 45–53.

Daly, R.A. (1934) *The Changing World of the Ice Age*. Yale University Press, New Haven, 271 pp.

Dansgaard, W., Johnsen, S.J., Clausen H.B., Dahl-Jensen, D., Gundestrup, N.S., Hammer, C.U., Hvidberg, C.S., Steffensen, J.P., Sveinbjörnsdottir, A.E., Jouzel, J. and Bond, G. (1993) Evidence for general instability of past climate from a 250-kyr ice-core record. *Nature*, **364**, 218–220.

Darienzo, M.E. and Peterson, C.D. (1990) Episodic tectonic subsidence of late Holocene salt marshes, northern Oregon central Cascadia margin. *Tectonics*, **9**, 1–22.

Darwin, C. (1846) *Geological Observations in South America*. London.

Davies, P.J. and Montaggioni, L. (1985) Reef growth and sea-level change: the environmental signature. *Proceedings of Fifth International Coral Reef Congress, Tahiti*, vol. 3, 477–515.

Davies, P.J., Marshall, J.F. and Hopley, D. (1985) Relationships between reef growth and sea level in the Great Barrier Reef. *Proceedings of Fifth International Coral Reef Congress*, Tahiti, vol. 3, 95–103.

Dawson, A.G. (1992) *Ice Age Earth*. Routledge, London, 293 pp.

Dawson, A.G., Long, D. and Smith, D.E. (1988) The Storegga Slides: evidence from eastern Scotland for a possible tsunami. *Marine Geology*, **88**, 271–276.

Dawson, A.G., Hindson, R., Andrade, C., Freitas, C., Parish, R. and Bateman, M. (1995) Tsunami sedimentation associated with the Lisbon earthquake of 1 November AD 1755: Boca do Rio, Algarve, Portugal. *The Holocene*, **5** (2), 209–215.

Degiovanni, C. (1973) *Etude de sédimentologie dynamique à la presqu'île de Sidi-Ferruch (ouest d'Alger)*. PhD in Marine Geology and Sedimentology, Université de Provence.

Denys, L. and Baeteman, C. (1995) Holocene evolution of relative sea level and local mean high water spring tides in Belgium – a first assessment. *Marine Geology*, **124**, 1–19.

Detrick, R.S. and Crough, S.T. (1978) Island subsidence, hot spots and lithospheric thinning. *Journal of Geophysical Research*, **83**, 1236–1244.

Devoy, R.J. (1985) The problem of a Late Quaternary landbridge between Britain and Ireland. *Quaternary Science Reviews*, **4**, 43–58.

Dillon, W.P. and Oldale, R.N. (1978) Late Quaternary sea-level curve: reinterpretation based on glaciotectonic influence. *Geology*, **6**, 56–60.

Doodson, A.T. and Warburg, H.D. (1941) *Admiralty Manual of Tides*. Stationery Office, London, 270 pp.

Douglas, B.C. (1991) Global sea level rise. *Journal of Geophysical Research*, **96** (C4), 6981–6992.

Douglas, B.C. (1992) Global sea level acceleration. *Journal of Geophysical Research*, **97** (C8), 12 699–12 706.

Douglas, B.C., Cheney, R.E. and Agreen, R.W. (1983) Eddy energy of the northwest Atlantic and Gulf of Mexico determined from GEOS 3 altimetry. *Journal of Geophysical Research*, **88** (C14), 9595–9603.

Dredge, L.A. and Nixon, F.M. (1992) *Glacial and Environmental Geology of Northeastern Manitoba*. Geological Survey of Canada, Memoir **432**, 80 pp.

Dubar, M. and Anthony, E.J. (1995) Holocene environmental change and river-mouth sedimentation in the Baie des Anges, French Riviera. *Quaternary Research*, **43**, 329–343.

Dumon, J.C., Froidefond, J.M., Gayet, J., Naudin, J.J., Peypouquet, J.P., Prud'homme, R., Saubade, A.M. and Turon, J.L. (1977) Evolution holocène de la couverture sédimentaire du proche plateau continental au Sud de Dakar (Sénégal). *Bulletin de la Société Géologique de France*, **19** (2), 219–234.

Edwards, R.L. (1995) Paleotopography of glacial-age ice sheets. *Science*, **267**, 536.

Edwards, R.L., Beck, J.W., Burr, G.S., Donahue, D.J., Chappell, J.M.A., Bloom, A.L., Druffel, E.R.M. and Taylor, F.W. (1993) A large drop in atmospheric $^{14}C/^{12}C$ and reduced melting in the Younger Dryas, documented with ^{230}Th ages of corals. *Science*, **260**, 962–968.

Einsele, G., Herm, D. and Schwartz, H.U. (1974) Sea level fluctuations during the past 6000 yr at the coast of Mauritania. *Quaternary Research*, **4**, 282–289.

Ellison, J.C. (1989) Pollen analysis of mangrove sediments as sea-level indicator: assessment from Tongatapu, Tonga. *Palaeogeography, Palaeoclimatology, Palaeoecology*, **74**, 327–341.

Ellison, J.C. and Stoddart, D.R. (1991) Mangrove ecosystem collapse during predicted sea-level rise: Holocene analogues implications. *Journal of Coastal Research*, **7** (1), 151–165.

Emery, K.O. (1980) Relative sea levels from tide-gauge records. *Proceedings of National Academy of Science USA*, **77** (12), 6968–6972.

Emery, K.O. and Aubrey, D.G. (1991) *Sea Levels, Land Levels, and Tide Gauges*. Springer-Verlag, New York, 237 pp.

Emery, K.O., Niino, H. and Sullivan B. (1971) Post-Pleistocene levels of the East China Sea. In K.K. Turekian (ed.) *The Late Cenozoic Glacial Ages*. Yale University Press, New Haven and London, pp. 381–390.

Etkins, R. and Epstein, E.S. (1982) The rise of global mean sea level as an indication of climatic change. *Science*, **215**, 287–289.

Fairbanks, R.G. (1989) A 17,000-year glacio-eustatic sea level record: influence of glacial melting rates on the Younger Dryas event and deep-ocean circulation. *Nature*, **342**, 637–642.

Fairbanks, R.G. and Matthews, R.K. (1978) The marine oxygen isotope record in Pleistocene coral, Barbados, West Indies. *Quaternary Research*, **10**, 181–196.

Fairbridge R.W. (1961) Eustatic changes in sea level. *Physics and Chemistry of the Earth*, **4**, 99–185.

Fairbridge, R.W. (1989) Crescendo events in sea-level changes. *Journal of Coastal Research*, **5** (1), ii–vi.

Fairbridge, R.W. and Krebs, O.A.Jr. (1962) Sea level and the Southern Oscillation. *Geophysical Journal*, **6**, 532–545.

Faure, H. (1980) Late Cenozoic vertical movements in Africa. In: N.A. Mörner (ed.) *Earth Rheology, Isostasy and Eustasy*. Wiley, Chichester, pp. 465–469.

Faure, H. (1990) Changes in the global continental reservoir of carbon *Global and Planetary Change*, **2**, 47–52.

Faure, H., Faure-Denard, L. and Liu, T. (1993) Introduction to Quaternary earth system changes. *Global and Planetary Change*, **7**, VII–IX.

Faure, H. and Elouard, P. (1967) Schéma des variations du niveau de l'océan Atlantique sur la côte de l'Ouest de l'Afrique depuis 40 000 ans. *Comptes Rendus de l'Académie des Sciences, Paris*, **265** (D), 784–787.

Faure, H., Fontes, J.C., Hébrard, L., Monteillet, J. and Pirazzoli, P.A. (1980) Geoidal change and shore-level tilt along Holocene estuaries: Sénégal River area, West Africa. *Science*, **210**, 421–423.

Fisher, D.A., Reeh, N. and Langley, K. (1985) Objective reconstructions of the late Wisconsinan Laurentide ice sheet and the significance of deformable beds. *Géographie Physique et Quaternaire*, **39** (3), 229–238.

Fjeldskaar, W. (1991) Geoidal-eustatic changes induced by the deglaciation of Fennoscandia. *Quaternary International*, **9**, 1–6.

Flemming, N.C. (1979–1980) Archaeological indicators of sea level. In: *Les Indicateurs de Niveaux Marins*. Oceanis, vol. 5, Fasc. Hors-Série, pp. 149–166.

Focke, J.M. (1978) Limestone cliff morphology on Curaçao (Netherlands Antilles), with special attention to the origin of notches and vermetid/coralline algal surf benches. *Zeitschrift für Geomorphologie*, **22**, 329–349.

Forman, S.L., Mann, D.H. and Miller, G.H. (1987) Late Weichselian and Holocene relative sea-level history of Bröggerhalvöya, Spitsbergen. *Quaternary Research*, **27**, 41–50.

Fujii, S. and Fuji, N. (1967) Postglacial sea level in the Japanese Islands. In: N. Ikebe (ed.) *Sea Level Changes and Crustal Movements of the Pacific During the Pliocene and Post-Pliocene Time. Journal of Geoscience, Osaka City University*, **10**, 43–51.

Fujii, S. and Mogi, A. (1970) On coasts and shelves in their mutual relations in Japan during the Quaternary. *Quaternaria*, **12**, 155–164.

Ganeko, Y. (1983) A 10' × 10' detailed gravimetric geoid around Japan. *Marine Geodesy*, **7**, 291–314.

Gasse, F., Téhet, R., Durand, A., Gibert, E. and Fontes, J.C. (1990) The arid–humid transition in the Sahara and the Sahel during the last deglaciation. *Nature*, **346** (6280), 141–146.

Gibb, J.G. (1986) A New Zealand regional Holocene eustatic sea-level curve and its application to determination of vertical tectonic movements. *Royal Society of New Zealand Bulletin*, **24**, 377–395.

Giresse, P., Malounguila-N'Ganga, D. and Barusseau, J.P. (1986) Submarine evidence of the successive shorefaces of the Holocene transgression off southern Gabon and Congo. *Journal of Coastal Research, Special Issue*, **1**, 61–71.

Godwin, H., Suggate, R.P. and Willis, E.H. (1958) Radiocarbon dating of the eustatic rise in ocean level. *Nature*, **181**, 1518–1519.

Gordillo, S., Bujalesky, G.G., Pirazzoli, P.A., Rabassa, J.O. and Saliège, J.F. (1992) Holocene raised beaches along the northern coast of the Beagle Channel, Tierra del Fuego, Argentina. *Palaeogeography, Palaeoclimatology, Palaeoecology*, **99**, 41–54.

Gornitz, V. (1993) Mean sea level changes in the recent past. In: R. Warrick et al. (eds.) *Climate and Sea Level Change: Observations, Projections and Implications*. Cambridge University Press, Cambridge, pp. 25–44.

Gornitz, V. and Lebedeff, S. (1987) Global sea-level changes during the past century. In: D. Nummedal et al. (eds.) *Sea-Level Fluctuation and Coastal Evolution*. Society of Economic Paleontologists and Mineralogists, Special Publication no. 41, pp. 3–16.

Gornitz, V., Lebedeff, S. and Hansen, J. (1982). Global sea level trend in the past century. *Science*, **215**, 1611–1614.

Gornitz, V., Rosenzweig, C. and Hillel, D. (1996) Effects of anthropogenic intervention in the land hydrologic cycle on global sea level rise. *Global and Planetary Change* (in press).

Grant, D.R. (1970) Recent coastal submergence of the Maritime Provinces, Canada. *Canadian Journal of Earth Science*, **7**, 676–689.

Grant, D.R. (1980) Quaternary sea-level change in Atlantic Canada as an indication of crustal delevelling. In: N.A. Mörner (ed.) *Earth Rheology, Isostasy and Eustasy*. Wiley, Chichester, pp. 201–214.

Gröger, M. and Plag, H.P. (1993) Estimations of a global sea level trend: limitations from the structure of the PSMSL global sea level data set. *Global and Planetary Change*, **8**, 161–169.

Grootes, P.M., Stuiver, M., White, J.W.C., Johnsen, S. and Jouzel, J. (1993). Comparison of the oxygen isotope records from the GISP2 and GRIP Greenland ice cores. *Nature*, **366**, 552–554.

Guilcher, A. (1961) Le "beach-rock" ou grès de plage. *Annales de Géographie*, **70**, 113–125.

Guilcher, A (1988) *Coral Reef Geomorphology*. Wiley, Chichester, 228 pp.

Guilcher, A., Berthois, L., Battistini, R. and Fourmanoir, P. (1958) Les récifs coralliens des îles Radama et de la baie Ramanetaka (côte nord-ouest de Madagascar), étude géomorphologique et sédimentologique. *Mémoires de l'Institut Scientifique de Madagascar*, F, **2**, 117–199.

Guilderson, T.P., Fairbanks, R.G. and Rubenstone, J.L. (1994) Tropical temperature variations since 20,000 years ago: modulating interhemispheric climate change. *Science*, **263**, 663–665.

Gutenberg, B. (1941) Changes in sea level, postglacial uplift, and mobility of the Earth's interior. *Bulletin of the Geological Society of America*, **52**, 721–772.

Hamilton, E.L. (1957) Marine geology of the southern Hawaiian ridge. *Bulletin of the Geological Society of America*, **68**, 1011–1026.

Haq, B.U., Hardembol, J. and Vail, P.R. (1987) Chronology of fluctuating sea levels since the Triassic. *Science*, **235**, 1156–1167.

Hebbein, D., Dokken, T., Andersen, E.S., Hald, M. and Elverhöl, A. (1994) Moisture supply for northern ice-sheet growth during the Last Glacial Maximum. *Nature*, **370**, 357–360.

Héquette, A. (1988) Vue récentes sur l'évolution du Svalbard au Quaternaire. *Revue de Géomorphologie Dynamique*, **37** (4), 129–141.

Heusser, C.J. and Rabassa, J. (1987) Cold climatic episode of Younger Dryas age in Tierra del Fuego. *Nature*, **328**, 609–611.

Hill, P.R., Mudie, P.J., Moran, K. and Blasco, S.M. (1985) A sea-level curve for the Canadian Beaufort shelf. *Canadian Journal of Earth Science*, **22**, 1383–1393.

Hillaire-Marcel, C. and Occhietti, S. (1980) Chronology, paleogeography and paleoclimatic significance of the late and post-glacial events in eastern Canada. *Zeitschrift für Geomorphologie*, **24**, 373–392.

Hinton, A.C. (1992) Palaeotidal changes within the area of the Wash during the Holocene. *Proceedings of the Geologists' Association*, **103**, 259–272.

Hinton, A.C. (1995) Holocene tides of The Wash, U.K.: the influence of water-depth and coastline-shape changes on the record of sea-level change. *Marine Geology*, **124**, 87–111.

Hoffman, J.S., Keyes, D. and Titus, J.G. (1983) *Projecting Future Sea Level Rise; Methodology, Estimates to the Year 2100, and Research Needs*. US Environmental Protection Agency, Washington, DC, 121 pp.

Hopkins, D.M. (1967) The Cenozoic history of Beringia-A Synthesis. In: *The Bering Land Bridges*. Stanford University Press, Stanford, CA, pp. 451–484.

Hopkins, D.M. (1983) Hard times in Beringia: A short note. In: P.M. Masters and N.C. Flemming (eds.), *Quaternary Coastlines and Marine Archaeology*. Academic Press, London, pp. 345–346.

Hopley, D. (1982) *The Geomorphology of the Great Barrier Reef*. Wiley, New York, 453 pp.

Hopley, D. (1986a) Beachrock as a sea-level indicator. In: O. Van de Plassche (ed.) *Sea-Level Research: A Manual for the Collection and Evaluation of Data*. Geo Books, Norwich, pp. 157–173.

Hopley, D. (1986b) Corals and reefs as indicators of paleo-sea levels. In: O. Van de Plassche (ed.) *Sea-Level Research: A Manual for the Collection and Evaluation of Data*. Geo Books, Norwich, pp. 195–228.

Hughes, T.J., Denton, G.H., Andersen, B.G., Schilling, D.H., Fastook, J.L. and Lingle, C.S. (1981) The last great ice sheets: a global view. In: G.H. Denton and T.J. Hughes (eds.) *The Last Great Ice Sheets*. Wiley, New York, pp. 275–318.

Ignatov Y.I., Kaplin, P.A., Lukyanova, S.A. and Solovieva, G.D. (1993) Evolution of the Caspian Sea coasts under conditions of sea-level rise: model for coastal change under increasing "greenhouse effect". *Journal of Coastal Research*, **9** (1), 104–111.

Intergovernmental Panel on Climate Change (1990) *Climate Change The IPCC Scientific Assessment*. Cambridge University Press, Cambridge, 365 pp.

Islebe, G.A., Hooghiemstra, H. and van der Borg, K. (1995) A cooling event during the Younger Dryas chron in Costa Rica. *Palaeogeography, Palaeoclimatology, Palaeoecology*, **117**, 73–80.

Jeftic, L., Milliman, J.D. and Sestini, G. (eds.) (1992) *Climatic Change and the Mediterranean*. Edward Arnold, London, 673 pp.

Jelgersma, S. (1961) *Holocene Sea-Level Changes in the Netherlands*. Mededelingen Geol. Stichting, vol. C-IV (7), 100 pp.

Jelgersma, S., Roep, T.B. and Beets, D.J. (1975) New data on sea-level changes in the Netherlands. *Guidebook INQUA Shoreline Meeting, France, Belgium, Netherlands, Germany*, Sept. 1975.

Johnsen, S.J., Clausen, H.B., Dansgaard, W., Fuhrer, K., Gunderstrup, N.S., Hammer, C.U., Iversen, P., Jouzel, J., Stauffer, B. and Steffensen, J.P. (1992) Irregular glacial interstadials recorded in a new Greenland ice core. *Nature*, **359**, 311–313.

Johnson, D.W. (1912) Fixité de la côte atlantique de l'Amérique du Nord. *Annales de Géographie*, **21**, 193–212.

Jongsma, D. (1970) Eustatic sea level changes in the Arafura Sea. *Nature*, **228**, 150–151.

Jouannic C., Taylor, F.W. and Bloom, A.L. (1982) Sur la surrection et la déformation d'un arc jeune: l'arc des Nouvelles-Hebrides. In: *Contribution à l'étude géodynamique du sud-ouest Pacifique*. Travaux et Documents ORSTOM, Paris, vol. 147, pp. 223–246.

Jouzel, J. (1994) Les enregistrements climatiques du Groënland et de l'Antarctique au cours du dernier cycle climatique. *Il Quaternario*, **7** (1b), 267–274.

Jouzel, J., Lorius, C., Johnsen, S. and Grootes, P. (1994) Climate instabilities: Greenland and Antarctic records. *Comptes Rendus de l'Académie des Sciences, Paris*, **319** (II), 65–77.

Julian, M., Lilienberg, D.A. and Nicod, J. (1987) Rôle des modifications des niveaux quaternaires des mers Caspienne, Noire et Méditerranée dans la formation du relief de leurs marges montagnardes. *Méditerranée*, **2–3**, 19–35.

Kaizuka, S., Matsuda, T., Nogami, M. and Yonekura, N. (1973) Quaternary tectonic and recent seismic crustal movements in the Arauco Peninsula and its environs, central Chile. *Geographical Report of Tokyo Metropolitan University*, **8**, 1–49.

Kalinin, G.P. and Klige, R.K. (1978). Variation in the world sea level. In: *World Water Balance and Water Resources of the Earth*. Studies and Reports in Hydrology, no. 25, Unesco, pp. 581–585.

Kallel, N., Labeyrie, L.D., Arnold, M., Okada, H., Dudley, W.C. and Duplessy, J.C. (1988) Evidence of cooling during the Younger Dryas in the western North Pacific. *Oceanologica Acta*, **11**, 369–375.

Kan, H, Hori, N., Nakashima, Y. and Ichikawa, K. (1995) The evolution of narrow reef flats at high-latitude in the Ryukyus. *Coral Reefs*, **14**, 123–130.

Kaplin, P., Pirazzoli, P.A., Pavlides, Y. and Badenkov, Y. (1986) Sea-level and environmental changes in shelf areas of the western Indian Ocean. *Journal of Coastal Research*, **2**, 363–367.

Kaplin, P.A. and Shcherbakov, F.A. (1986) Reconstruction of shelf environments during the late Quaternary. *Journal of Coastal Research, Special Issue*, **1**, 95–98.

Kawana, T. and Pirazzoli, P.A. (1985) Holocene coastline changes and seismic uplift in Okinawa Island, the Ryukyus, Japan. *Zeitschrift für Geomorphologie, Supplement*, **57**, 11–31.

Kawana, T. and Pirazzoli, P.A. (1990) Re-examination of the Holocene emerged shorelines in Irabu and Shimoji Islands, the South Ryukyus, Japan. *The Quaternary Research* (Tokyo), **28**, 419–426.

Kaye, C.A. (1959) *Shoreline Features and Quaternary Shoreline Changes Puerto Rico*. Geological Survey, Professional Paper no. 317-B, 136 pp.

Kaye, C.A. and Barghoorn, E.S. (1964) Late Quaternary sea-level change and crustal rise at Boston, Massachusetts, with notes on autocompaction of peat. *Bulletin of the Geological Society of America*, **75**, 63–80.

Keeling, C.D., Whorf, T.P., Wahlen, M. and Van der Plicht, J. (1995) Interannual extremes in the rate of rise of atmospheric carbon dioxide since 1980. *Nature*, **375**, 666–670.

Keffer, T., Martinson, D.G. and Corliss, B.H. (1988) The position of the Gulf Stream during Quaternary glaciations. *Science*, **241**, 440–442.

Kelley, J.T., Belknap, D.F., Jacobson, J.L. Jr and Jacobson, H.A. (1988) The morphology and origin of salt marshes along the glaciated coastline of Maine, USA. *Journal of Coastal Research*, **4**, 649–665.

Kirk, R.M. (1977) Rates and forms of erosion on intertidal platforms at Kaikoura Peninsula, South Island, New Zealand. *New Zealand Journal of Geology and Geophysics*, **20**, 571–613.

Kjemperud, A. (1986) Late Weichselian and Holocene shoreline displacement in the Trondheimsfjord area, central Norway. *Boreas*, **15**, 61–82.

Klige, R.K. and Dobrovolsky, S.G. (1988) The ocean level and models simulating its oscillations. *Journal of Coastal Research*, **4**, 273–278.

Kraft, J.C., Rapp, G.Jr., Szemler, G.J., Tziavos, C. and Kase, E.W. (1987) The Pass at Thermopylae, Greece. *Journal of Field Archaeology*, **14**, 181–198.

Kudrass, H.R., Erlenkeuser, H., Vollbrecht, R. and Weiss, W. (1991) Global nature of the Younger Dryas cooling event inferred from oxygen isotope data from Sulu Sea cores. *Nature*, **349**, 406–409.

Labeyrie, J., Lalou, C. and Delibrias, G. (1969) Etude des transgressions marines sur l'atoll de Mururoa par la datation des différents niveaux de corail. *Cahiers du Pacifique*, **13**, 59–68.

Labeyrie, J., Lalou, C., Monaco, A. and Thommeret, J. (1976) Chronologie des niveaux eustatiques sur la côte du Roussillon de –33 000 ans B.P. à nos jours. *Comptes Rendus de l'Académie des Sciences, Paris*, **282** (D), 349–352.

Laborel, J. (1987) Marine biogenic constructions in the Mediterranean – A review. *Sci. Rep. Port-Cros National Park, France*, **13**, 97–126.

Laborel, J. and Laborel-Deguen, F. (1994) Biological indicators of relative sea-level variations and of co-seismic displacements in the Mediterranean region. *Journal of Coastal Research*, **10**, 395–415.

Laborel, J., Morhange, C., Lafont, R., Le Campion, J., Laborel-Deguen, F. and Sartoretto, S. (1994) Biological evidence of sea-level rise during the last 4500 years on the rocky coasts of continental southwestern France and Corsica. *Marine Geology*, **120**, 203–223.

Lambeck, K. (1993a) Glacial rebound of the British Isles – I. Preliminary model results. *Geophysical Journal International*, **115**, 941–959.

Lambeck, K. (1993b) Glacial rebound of the British Isles – II. A high-resolution, high-precision model. *Geophysical Journal International*, **115**, 960–990.

Lambeck, K. (1993c) Glacial rebound and sea-level change: an example of a relationship between mantle and surface processes. *Tectonophysics*, **223**, 15–37.

Lambeck, K. (1995) Late Pleistocene and Holocene sea-level change in Greece and southwestern Turkey: a separation of eustatic, isostatic, and tectonic contributions. *Geophysical Journal International*, **122**, 1022–1044.

Lambeck, K. (1996) Sea-level change along the French Atlantic and Channel coasts since the time of the last glacial maximum. *Palaeogeography, Palaeoclimatology, Palaeoecology* (in press).

Lambeck, K. and Johnston, P. (1995) Land subsidence and sea-level change: Contributions from the melting of the last great ice sheets and the isostatic adjustment of the Earth. In: F.B. Barends et al. (eds.) *Land Subsidence.* Proceedings of the Fifth International Symposium on Land Subsidence, The Hague, Netherlands, 16–20 October 1995. Balkema, Rotterdam, pp. 3–18.

Le Campion-Alsumard, T. (1979–1980) Les végétaux perforants en tant qu'indicateurs paléobathymétriques. In: NIVMER (ed.) *Les Indicateurs de Niveaux Marins.* Oceanis, vol. 5, Fasc. Hors-Série, pp. 259–264.

Lehman, S.J. and Keigwin, L.D. (1992) Sudden changes in North Atlantic circulation during the last deglaciation. *Nature*, **356**, 757–762.

Liew, P.M., Pirazzoli, P.A., Hsieh, M.L., Arnold, M., Barusseau, J.P., Fontugne, M. and Giresse, P. (1993) Holocene tectonic uplift deduced from elevated shorelines, eastern Coastal Range of Taiwan. *Tectonophysics*, **222**, 55–68.

Lisitzin, E. (1958) Le niveau moyen de la mer. *Bulletin d'Informations du Comité Central d'Océanographie et d'Etude des Côtes (COEC)*, **10**, 254–262.

Lisitzin, E. (1965) The mean sea level of the world ocean. *Societas Scientiarum Fennica, Commentationes Physico-Mathematicae*, **30** (7), 5–35.

Lisitzin, E. (1974) *Sea-Level Changes.* Elsevier, New York, Oceanography Series no. 8, 286 pp.

Locat, J. (1977) L'émersion des terres dans la région de Baie-des-Sables/Trois-Pistoles, Québec. *Géographie Physique et Quaternaire*, **31**, 297–306.

Long, D., Smith, D.E. and Dawson, A.G. (1989) A Holocene tsunami deposit in eastern Scotland. *Journal of Quaternary Science*, **4**, 61–66.

Lorius, C., Jouzel, J., Ritz, C., Merlivat, L., Barkov, N.I., Korotkevich, Y.S. and Kotlyakov, V.M. (1985). A 150,000-year climatic record from Antarctic ice. *Nature*, **316**, 591–596.

Lorius, C., Barkov, N.I., Jouzel, J., Korotkevitch, Y.S., Kotlyakov, V.M. and Raynaud, D. (1988) Antarctic ice core: CO_2 and climate change over the last climate cycle. *EOS*, June 28, 681–684.

Lowell, T.V., Heusser, C.J., Andersen, B.G., Moreno, P.I., Hauser, A., Heusser, L.E., Schlüchter, C., Marchant, D.R. and Denton, G.H. (1995) Interhemispheric correlation of late Pleistocene glacial events. *Science*, **269**, 1541–1549.

Macintyre, I.G., Pilkey, O.H. and Stuckenrath, R. (1978) Relict oysters on the United States Atlantic continental shelf: A reconsideration of their usefulness in understanding late Quaternary sea-level history. *Geological Society of America Bulletin*, **89**, 277–282.

Madeyska, T., Soffer, O. and Kurenkova, E.I. (1992) Human occupation. In: B. Frenzel, M. Pécsi and A.A. Velichko (eds.) *Atlas of Paleoclimates and Paleoenvironments of the Northern Hemisphere.* Geographical Research Institute, Hungarian Academy of Sciences, Budapest, plate 63.

Maillet, B. de (1748) *Telliamed ou Entretiens d'un philosophe indien avec un missionnaire français sur la diminution de la mer.* Amsterdam.

Manfredi, E. (1746) *De Aucta Maris Altitudine.* De Bonon, Scient. Art. Inst. Acad. Commentarii, vol. I, no. 2, pp. 1–19 (quoted by Suess, 1885).

Maragos, J.E., Baines, G.B.K. and Beveridge, P.J. (1973) Tropical cyclone Bebe creates a new land formation on Funafuti Atoll. *Science*, **181**, 1161–1164.

Markovic-Marjanovic, J. (1971) Coast lines, Pleistocene sediments and fauna of the eastern part of Adriatic in Yugoslavia. *Quaternaria*, **XV**, 187–195.

Marmer, H.A. (1943) Tide observations at Baltimore and the problem of coastal stability. *Geographical Review*, **33**, 620–629.

Marthinussen, M. (1961) C_{14}-datings referring to shore lines, transgressions, and glacial substages in Northern Norway. *Norges Geologiske Undersokelse.*, **215**, 37–67.

Martin, L. and Delibrias, G. (1972) Schéma des variations du niveau de la mer en

Côte-d'Ivoire depuis 25 000 ans. *Comptes Rendus de l'Académie des Sciences, Paris,* **274** (D), 2848–2851.

Masters, P.M and Flemming, N.C. (eds.) (1983) *Quaternary Coastlines and Marine Archaeology.* Academic Press, London, 641 pp.

Mathewes, R.W. and Clague, J.J. (1994) Detection of large prehistoric earthquakes in the Pacific Northwest by microfossil analysis. *Science,* **264,** 688–691.

Matthews, R.K. (1990) Quaternary sea-level change. In: *Sea-Level Change.* Studies in Geophysics, National Academy Press, Washington, DC, pp. 88–103.

Maul, G.A. (ed.) (1993) *Climatic Change in the Intra-Americas Sea.* Edward Arnold, London, 389 pp.

Maul, G.A., Hendry, M.D. and Pirazzoli, P.A. (1996) Sea level, tide and tsunamis. In: G.A. Maul (ed.) *Small Islands: Marine Science and Sustainable Development.* Coastal and Estuarine Studies, American Geophysical Union (in press).

Mayewski, P.A., Meeker, L.D., Whitlow, S., Twickler, M.S., Morrison, M.C., Alley, R.B., Bloomfield, P. and Taylor, K. (1993) The atmosphere during the Younger Dryas. *Science,* **261,** 195–197.

McManus, D.A. and Creager, J.S. (1984) Sea-level data for parts of the Bering-Chukchi shelves of Beringia from 19,000 to 10,000 14C yr B.P. *Quaternary Research,* **21,** 317–325.

McManus, D.A., Creager, J.S., Echols, R.J. and Holmes, M.L. (1983) The Holocene transgression on the Arctic Flank of Beringia: Chukchi Valley to Chukchi Estuary to Chukchi Sea. In: P.M. Masters and N.C. Flemming (eds.) *Quaternary Coastlines and Marine Archaeology.* Academic Press, London, pp. 365–388.

McMaster, R., Lachance, T.P. and Ashraf, A. (1970) Continental shelf geomorphic features off Portuguese Guinea, Guinea, and Sierra Leone (West Africa). *Marine Geology,* **9,** 203–213.

McNutt, M. and Menard, H.W. (1978) Lithospheric flexure and uplifted atolls. *Journal of Geophysical Research,* **83,** 1206–1212.

Meier, M.F. (1984) Contribution of small glaciers to global sea level. *Science,* **226** (468), 1418–1421.

Meighan, C.M. (1983) Early man in the New World. In: P.M. Masters and N.C. Flemming (eds.) *Quaternary Coastlines and Marine Archaeology.* Academic Press, London, pp. 441–461.

Menard, H.W. (1973) Depth anomalies and the bobbing motion of drifting islands. *Journal of Geophysical Research,* **78** (23), 5128–5137.

Menard, H.W. and Ladd, H.P. (1963) Oceanic islands, sea-mounts, guyots and atolls. In: M.N. Hill (ed.) *The Sea.* Wiley Interscience, New York, vol. 3, pp. 365–387.

Miller, R.L. and Zeigler, J.M. (1958) A model relating dynamics and sediment pattern in equilibrium in the region of shoaling waves, breaker zone and foreshore. *Journal of Geology,* **66,** 417–441.

Milliman, J.D. and Emery, K.O. (1968) Sea levels during the past 35,000 years. *Science,* **162,** 1121–1123.

Mitrovica, J.X. and Peltier, W.R. (1991) On postglacial geoid subsidence over the equatorial oceans. *Journal of Geophysical Research,* **96** (B12), 20 053–20 071.

Montaggioni, L.F. (1988) Holocene reef growth history in mid-plate high volcanic islands. In: *Proceedings of 6th International Coral Reef Symposium, Australia,* vol. 3, pp. 455–460.

Montaggioni, L.F. and Pirazzoli, P.A. (1984) The significance of exposed coral conglomerates from French Polynesia (Pacific Ocean) as indicators of recent relative sea-level changes. *Coral Reefs,* **3,** 29–42.

Mook, W.G. and Van de Plassche, O. (1986) Radiocarbon dating. In: O. Van de Plassche (ed.) *Sea-Level Research: a Manual for the Collection and Evaluation of Data*. Geo Books, Norwich, pp. 525–560.

Moore, J.G. (1971) Relationship between subsidence and volcanic load, Hawaii. *Bulletin Volcanologique*, **34**, 562–576.

Morhange, C. (1994) *La Mobilité Récente des Littoraux Provençaux: Eléments d'Analyse Géomorphologique*. Thèse de doctorat en Géographie Physique, Université de Provence, Aix-en-Provence.

Mörner, N.A. (1969) *The Late Quaternary History of the Kattegatt Sea and the Swedish West Coast, Deglaciation, Shorelevel Displacement, Chronology, Isostasy and Eustasy*. Sveriges Geologiska Undersökning, C, vol. 640, pp. 1–487.

Mörner, N.A. (1973) Eustatic changes during the last 300 years. *Palaeogeography, Palaeoclimatology, Palaeoecology*, **13**, 1–14.

Mörner, N.A. (1976) Eustatic changes during the last 8,000 years in view of radiocarbon calibration and new information from the Kattegatt region and other northwestern European coastal areas. *Palaeogeography, Palaeoclimatology, Palaeoecology*, **19**, 63–85.

Mörner, N.A. (1979) The Fennoscandian uplift and Late Cenozoic geodynamics: geological evidence. *Geojournal*, **3** (3), 287–318.

Mörner, N.A. (1979-1980) The northwest European "sea-level laboratory" and regional Holocene eustasy. *Palaeogeography, Palaeoclimatology, Palaeoecology*, **29**, 281–300.

Mörner, N.A. (1980) The Fennoscandian uplift: geological data and their geodynamical implication. In: N.A. Mörner (ed.) *Earth Rheology, Isostasy and Eustasy*. Wiley, Chichester, pp. 251–284.

Mörner, N.A. (1986) The concept of eustasy: a redefinition. *Journal of Coastal Research, Special Issue*, **1**, 49–51.

Mörner, N.A. (1994) Recorded sea level variability in the Holocene and expected future changes. *Bulletin of the INQUA Neotectonic Commission*, **17**, 48–53.

Nakada, M. (1986) Holocene sea levels in oceanic islands: implications for the rheological structure of the earth's mantle. *Tectonophysics*, **121**, 263–276.

Nakada, M. and Lambeck, K. (1989) Late Pleistocene and Holocene sea-level change in the Australian region and mantle rheology. *Geophysical Journal*, **96**, 497–517.

Nakada, M., Yonekura, N. and Lambeck, K. (1991) Late Pleistocene and Holocene sea-level changes in Japan: implications for tectonic histories and mantle rheology. *Palaeogeography, Palaeoclimatology, Palaeoecology*, **85**, 107–122.

Nakata, T., Koba, M., Jo, W., Imaizumi, T., Matsumoto, H. and Suganuma, T. (1979) Holocene marine terraces and seismic crustal movement. *Science Reports of the Tohoku University, 7th Ser. (Geography)*, **29**, 195–204.

Nerem, R.S. (1995a) Global mean sea level variations from TOPEX/POSEIDON altimeter data. *Science*, **268**, 708–710.

Nerem, R.S. (1995b) Measuring global mean sea level variations using TOPEX/POSEIDON altimeter data. *Journal of Geophysical Research*, **100** (C12), 25 135–25 151.

Newman, W.S., Fairbridge, R.W. and March, S. (1971) Marginal subsidence of glaciated areas: United States, Baltic and North Seas. In: M. Ters (ed.) *Etudes sur le Quaternaire dans le Monde*. VIII Congress INQUA Paris, 1969, pp. 795–801.

Newman, W.S., Cinquemani, L.J., Pardi, R.R. and Marcus, L.F. (1980) Holocene delevelling of the United States east coast. In: N.A. Mörner (ed.) *Earth Rheology, Isostasy and Eustasy*. Wiley, Chichester, pp. 449–463.

Nunn, P.D. (1994) *Oceanic Islands*. Blackwell, Oxford, 413 pp.

Oerlemans, J. (1993) Possible changes in the mass balance of the Greenland and Antarctic ice sheets and their effects on sea level. In: R. Warrick et al. (eds.) *Climate and Sea Level Change*. Cambridge University Press, Cambridge, pp. 144–161.

Ortlieb, L., Barrientos, S., Ruegg, J.C., Guzman, N. and Lavenu, A. (1995) Coseismic coastal uplift during the 1995 Antofagasta earthquake. *2nd Annual Meeting of IGCP Project 367, Antofagasta, Chile, 19–28 Nov. 1995, Abstracts*, p. 83.

Ota, Y. (1985) Marine terraces and active faults in Japan with special reference to coseismic events. In: M. Morisawa and J.T. Hack (eds.) *Tectonic Geomorphology*. Allen and Unwin, Boston, pp. 345–366.

Ota, Y., Pirazzoli, P.A., Kawana, T. and Moriwaki, H. (1985) Late Holocene coastal morphology and sea-level records on three small islands, the South Ryukyus, Japan. *Geographical Review of Japan (Ser. B)*, **58** (2), 185–194.

Ota, Y., Miyauchi, T and Hull, A. G. (1990) Holocene marine terraces at Aramoana and Pourerere, eastern North Island, New Zealand. *New Zealand Journal of Geology and Geophysics*, **33**, 541–546.

Ota, Y., Hull, A.G. and Berryman, R. (1991) Coseismic uplift of Holocene marine terraces in the Pakarae River area, eastern North Island, New Zealand. *Quaternary Research*, **35**, 331–346.

Palmer, A.J.M. and Abbott, W.H. (1986) Diatoms as indicators of sea-level change. In: O. Van de Plassche (ed.) *Sea-Level Research: a Manual for the Collection and Evaluation of Data*. Geo Books, Norwich, pp. 457–487.

Parrilla, G., Laví, A., Bryden, H., García, M. and Millard, R. (1994) Rising temperatures in subtropical North Atlantic Ocean over the past 35 years. *Nature*, **369**, 48–51.

Paskoff, R. (1970) *Le Chili Semi-Aride: Recherches Géomorphologiques*. Biscaye Frères, Bordeaux, 420 pp.

Paskoff, R., Hurst, H. and Rakob, F. (1985) Position du niveau de la mer et déplacement de la ligne de rivage à Carthage (Tunisie) dans l'Antiquité. *Comptes Rendus de l'Académie des Sciences*, Paris, **300**, II (13), 613–618.

Pattullo, J., Munk, W., Revelle, R. and Strong, E. (1955) The seasonal oscillation in sea level. *Journal of Marine Research*, **14**, 88–156.

Peltier, W.R. (1974) The impulse response of a Maxwell earth. *Reviews of Geophysics and Space Physics*, **12**, 649–669.

Peltier, W.R. (1976) Glacial isostatic adjustments II: The inverse problem. *Geophysical Journal of the Royal Astronomical Society*, **46**, 669–705.

Peltier, W.R. (1990) Glacial isostatic adjustment and relative sea-level change. In: *Sea-Level Change*. Geophysics Study Committee, National Research Council, National Academy Press, Washington, DC, pp. 73–87.

Peltier, W.R. (1994) Ice age paleotopography. *Science*, **265**, 195–201.

Peltier, W.R. and Andrews, J.T. (1976) Glacial-isostatic adjustment I: the forward problem. *Geophysical Journal of the Royal Astronomical Society*, **46**, 605–646.

Peltier, W.R. and Tushingham, A.M. (1989). Global sea level rise and greenhouse effect: might they be connected? *Science*, **244**, 806–810.

Pérès, J.M. and Picard, J. (1964) *Nouveau manuel de bionomie benthique de la Mer Méditerranée*. Recueil des Travaux de la Station Marine d'Endoume, Bull. 31, fasc. 47, pp. 1–137.

Petersen, K.S. (1986) Marine molluscs as indicators of former sea-level stands. In: O. Van de Plassche (ed.) *Sea-Level Research: a Manual for the Collection and Evaluation of Data*. Geo Books, Norwich, pp. 129–155.

Petit-Maire, N. (1986) Palaeoclimates in the Sahara of Mali. *Episodes*, **9** (1), 7–16.

190 *References*

Pinot, J.P. (1968) Littoraux wurmiens submergés à l'ouest de Belle-Ile. *Bulletin de l'Association Française pour l'Etude du Quaternaire*, **3**, 197–216.

Pirazzoli, P.A. (1974) Dati storici sul medio mare a Venezia. *Atti della Accademia delle Scienze dell'Istituto di Bologna, Rendiconti*, **13** (1), 125–148.

Pirazzoli, P.A. (1975) Sulle ampiezze di marea nella laguna di Venezia. *Atti della Accademia delle Scienze dell'Istituto di Bologna, Rendiconti*, **13** (2), 124–155.

Pirazzoli, P.A. (1976) Sea level variations in the northwest Mediterranean during Roman times. *Science*, **194**, 519–521.

Pirazzoli, P.A. (1986a) Marine notches. In: O. Van de Plassche (ed.) *Sea-Level Research: a Manual for the Collection and Evaluation of Data*. Geo Books, Norwich, pp. 361–400.

Pirazzoli, P.A. (1986b) Secular trends of relative sea-level (RSL) changes indicated by tide-gauge records. *Journal of Coastal Research, Special Issue*, **1**, 1–26.

Pirazzoli, P.A. (1986c) Shelf areas of the western Indian Ocean-Late Quaternary evolution. *Episodes*, **9** (1), 30–31.

Pirazzoli, P.A. (1987a) A reconnaissance and geomorphological survey of Temoe Atoll, Gambier Islands (South Pacific). *Journal of Coastal Research*, **3**, 307–323.

Pirazzoli, P.A. (1987b) Recent sea-level changes and related engineering problems in the Lagoon of Venice. *Progress in Oceanography*, **18**, 323–346.

Pirazzoli, P.A. (1988) Sea-level changes and crustal movements in the Hellenic arc (Greece) – The contribution of archaeological and historical data. In: A. Raban (ed.) *Archaeology of Coastal Changes*. BAR International Series no. 404, Oxford, pp. 157–184.

Pirazzoli, P.A. (1989) Recent sea-level changes in the North Atlantic. In: D.B. Scott, P.A. Pirazzoli and C.A. Honig (eds.) *Late Quaternary Sea-Level Correlation and Applications*. Kluwer, Dordrecht NATO ASI Series C, vol. 256, pp. 153–167.

Pirazzoli, P.A. (1991) *World Atlas of Holocene Sea-Level Changes*. Elsevier, Amsterdam, Oceanography Series, vol. 58, 300 pp.

Pirazzoli, P.A. (1993) Global sea-level changes and their measurements. *Global and Planetary Change*, **8**, 135–148.

Pirazzoli, P.A. (1995) Tectonic shorelines. In: R.W.G. Carter and C.D. Woodroffe (eds.) *Coastal Evolution: Late Quaternary Shoreline Morphodynamics*. Cambridge University Press, Cambridge, pp. 451–476.

Pirazzoli, P.A. and Kawana, T. (1986) Détermination de mouvements crustaux quaternaires d'après la déformation des anciens rivages dans les îles Ryukyu, Japon. *Revue de Géologie Dynamique et de Géographie Physique*, **27**, 269–278.

Pirazzoli, P.A. and Montaggioni, L.F. (1985) Lithospheric deformation in French Polynesia (Pacific Ocean) as deduced from Quaternary shorelines. *Proceedings Fifth International Coral Reef Congress, Tahiti*, vol. 3, pp. 195–200.

Pirazzoli, P.A. and Montaggioni, L.F. (1988) Holocene sea-level changes in French Polynesia. *Palaeogeography, Palaeoclimatology, Palaeoecology*, **68**, 153–175.

Pirazzoli, P.A., Thommeret, J., Thommeret, Y., Laborel, J. and Montaggioni, L.F. (1982) Crustal block movements from Holocene shorelines: Crete and Antikythira (Greece). *Tectonophysics*, **86**, 27–43.

Pirazzoli, P.A., Brousse, R., Delibrias, G., Montaggioni, L.F., Sachet M.H., Salvat, B. and Sinoto, Y.H. (1985a) Leeward Islands: Maupiti, Tupai, Bora Bora, Huahine, Society Archipelago. *Proceedings Fifth International Coral Reef Congress, Tahiti*, vol. 1, pp. 17–72.

Pirazzoli, P.A., Delibrias, G., Kawana, T. and Yamaguchi, T. (1985b) The use of barnacles to measure and date relative sea-level changes in the Ryukyu Islands, Japan. *Palaeogeography, Palaeoclimatology, Palaeoecology*, **49**, 161–174.

Pirazzoli, P.A., Koba, M., Montaggioni, L.F. and Person, A. (1988a) Anaa (Tuamotu Islands, central Pacific): an incipient rising atoll? *Marine Geology*, **82**, 261–269.

Pirazzoli, P.A., Montaggioni, L.F., Salvat, B. and Faure, G. (1988b) Late Holocene sea level indicators from twelve atolls in the central and eastern Tuamotus (Pacific Ocean). *Coral Reefs*, **7**, 57–68.

Pirazzoli, P.A., Montaggioni, L.F., Saliège, J.F., Segonzac, G., Thommeret, Y. and Vergnaud-Grazzini, C. (1989) Crustal block movements from Holocene shorelines: Rhodes Island (Greece). *Tectonophysics*, **170**, 89–114.

Pirazzoli, P.A., Ausseil-Badie, J., Giresse, P., Hadjidaki, E. and Arnold, M. (1992) Historical environmental changes at Phalasarna harbor, West Crete. *Geoarchaeology*, **7**, 371–392.

Pirazzoli, P.A., Arnold, M., Giresse, P., Hsieh, M.L. and Liew, P.M. (1993) Marine deposits of late glacial times exposed by tectonic uplift on the east coast of Taiwan. *Marine Geology*, **110**, 1–6.

Pirazzoli, P.A., Laborel, J. and Stiros, S.C. (1996a) Earthquake clustering in the Eastern Mediterranean during historical times. *Journal of Geophysical Research*, **101** (B3), 6083–6097.

Pirazzoli, P.A., Laborel, J. and Stiros, S.C. (1996b) Coastal indicators of rapid uplift and subsidence: examples from Crete and other Eastern Mediterranean sites. *Zeitschrift für Geomorphologie, Supplement*, **102**, 21–35.

Plafker, G. and Rubin, M. (1967) Vertical tectonic displacements in south-central Alaska during and prior to the great 1964 earthquake. *Journal of Geosciences, Osaka City University*, **10**, 53–66.

Polli, S. (1952) Gli attuali movimenti verticali delle coste continentali. *Annali di Geofisica*, **5** (4), 597–602.

Porter, S.C. (1979) Hawaiian glacial ages. *Quaternary Research*, **12**, 161–187.

Pugh, D.T. (1987) *Tides, Surges and Mean Sea Level*. Wiley, Chichester, 472 pp.

Quinlan, G. and Beaumont, C. (1981) A comparison of observed and theoretical postglacial relative sea level in Atlantic Canada. *Canadian Journal of Earth Science*, **18**, 1146–1163.

Quinlan, G. and Beaumont, C. (1982) The deglaciation of Atlantic Canada as reconstructed from the postglacial relative sea-level record. *Canadian Journal of Earth Science*, **19**, 2232–2246.

Raynaud, D. and Barnola, J.M. (1985) An Antarctic ice core reveals atmospheric CO_2 variations over the past two centuries. *Nature*, **315**, 309–311.

Read, J.F. and Gould, W.J. (1992) Cooling and freshening of the subpolar North Atlantic Ocean since the 1960s. *Nature*, **360**, 55–57.

Reeves, B.O.K. (1983) Bergs, barriers and Beringia: reflections on the peopling of the new world. In: P.M. Masters and N.C. Flemming (eds.) *Quaternary Coastlines and Marine Archaeology*. Academic Press, London, pp. 389–411.

Rind, D., Peteet, D., Broecker, W., Mcintyre, A. and Ruddiman, W. (1986) Impact of cold North Atlantic sea surface temperatures on climate implications for the Younger Dryas cooling. *Climate Dynamics*, **1**, 3–33.

Robin, G. de Q. (1986) Changing the sea level. In: B. Bolin et al. (eds.) *The Greenhouse Effect. Climatic Change and Ecosystems*. SCOPE 29, Wiley, New York, pp. 323–359.

Roemmich, D. (1992) Ocean warming and sea level rise along the southwest U.S. coast. *Science*, **257**, 373–375.

Roemmich, D. and Wunsch, C. (1984) Apparent changes in the climatic state of the deep North Atlantic Ocean. *Nature*, **307**, 447–450.

Ruddiman, W.F. and McIntyre, A. (1981) The mode and mechanism of the last deglaciation: oceanic evidence. *Quaternary Research*, **16**, 125–134.

Ruddiman, W.F. and Raymo, M.E. (1988) Northern Hemisphere climate regimes during the past 3 Ma: possible tectonic connections. *Philosophical Transactions of the Royal Society of London*, B, **318**, 411–430.

Russell, G. (1991) Vertical distribution. In: A.C. Mathieson and P.H. Nienhuis (eds.), *Intertidal and Littoral Ecosystems*. Elsevier, Amsterdam, pp. 43–65.

Russell, R.J. and McIntyre, W.G. (1965) Southern Hemisphere beach rock. *Geographical Review*, **55**, 17–45.

Sabadini, R., Doglioni, C. and Yuen, D.A. (1990) Eustatic sea level fluctuations induced by polar wander. *Nature*, **345**, 708–710.

Sahagian, D.L., Schwartz, F.W. and Jacobs, D.K. (1994) Direct anthropogenic contributions to sea level rise in the twentieth century. *Nature*, **367**, 54–57.

Salvigsen, O. (1981) Radiocarbon dated raised beaches in Kong Karls Land, Svalbard, and their consequences for the glacial history of the Barents Sea area. *Geografiska Annaler, A*, **63** (3–4), 283–291.

Sandwell, D.T. (1991) *Geophysical application of satellite altimetry. Review of Geophysics, Supplement*, US National Report to IUGG 1987–1990, pp. 132–137.

Sartoretto, S., Collina-Girard, J., Laborel, J. and Morhange, C. (1995) Quand la grotte Cosquer a-t-elle été fermée par la montée des eaux? *Méditerranée*, **82** (3–4), 21–24.

Savage, J.C. and Plafker, G. (1991) Tide gage measurements of uplift along the south coast of Alaska. *Journal of Geophysical Research*, **96**, 4325–4335.

Schmiedt, G. (1972) *Il Livello Antico del Mar Tirreno*. Olschki, Firenze, 323 pp.

Scoffin, T.P. (1993) The geological effects of hurricanes on coral reefs and the interpretation of storm deposits. *Coral Reefs*, **12**, 203–221.

Scoffin, T.P. and Stoddart, D.R. (1978) The nature and significance of microatolls. *Philosophical Transactions of the Royal Society of London*, B, **284**, 99–122.

Scott, D.B. and Medioli, F.S. (1980) Post-glacial emergence curves in the Maritimes determined from marine sediments in raised basins. In: *Canadian Coastal Conference 1980, Proceedings*, Burlington, Ontario, pp. 428–446.

Scott, D.B. and Medioli, F.S. (1986) Foraminifera as sea-level indicators. In: O. Van de Plassche (ed.) *Sea-Level Research: a Manual for the Collection and Evaluation of Data*. Geo Books, Norwich, pp. 435–456.

Šegota, T. (1982–1983) Paleogeografske promjene u Jadranskom moru od Virmskog maksimuma do Danas. *Radovi Geografskog Odjela, Zagreb*, **17–18**, 11–15.

Serebryanny, L.R. (1982) Postglacial Black-Sea coast fluctuations and their comparison with the glacial history of the Caucasian high mountain region. In: P.A. Kaplin et al. (eds.) *Sea and Oceanic Level Fluctuations for 15000 Years* (in Russian). Nauka, Moscow, pp. 161–167.

Shackleton, N.J. (1987) Oxygen isotopes, ice volume and sea level. *Quaternary Science Reviews*, **6**, 183–190.

Shackleton, N.J. (1989) Deep trouble for climate change. *Nature*, **342**, 616–617.

Shackleton, N.J. and Opdyke, N.D. (1973) Oxygen isotope and paleomagnetic stratigraphy of equatorial Pacific core V28-238: oxygen isotope temperature and ice volumes on a 10^5 year and 10^6 year scale. *Quaternary Research*, **3**, 39–55.

Shackleton, N.J., Berger, A. and Peltier, W.R. (1990) An alternative astronomical calibration on the Lower Pleistocene time scale based on ODP Site 677. *Transactions of the Royal Society, London*, B, **318**, 679–688.

Shennan, I. (1982a) Interpretation of Flandrian sea-level data from the Fenland, England. *Proceedings of the Geologists' Association*, **93**, 53–63.

Shennan, I. (1982b) Problems of correlating Flandrian sea-level changes and climate. In: A.F. Harding (ed.) *Climatic Change in Later Prehistory*. Edinburgh University Press, Edinburgh, pp. 52–67.

Shennan, I. (1986) Flandrian sea-level changes in the Fenland. II: Tendencies of sea-level movement, altitudinal changes, and local and regional factors. *Journal of Quaternary Science*, **1**, 155–179.

Shennan, I. (1989) Holocene sea-level changes and crustal movements in the North Sea region: an experiment with regional eustasy. In: D.B. Scott, P.A. Pirazzoli and C.A. Honig (eds.) *Late Quaternary Sea-Level Correlation and Applications*. Kluwer, Dordrecht, NATO ASI Series C, vol. 256, pp. 1–25.

Shennan, I. and Tooley, M.J. (1987) Conspectus of fundamental and strategic research on sea-level changes. In: M.J. Tooley and I. Shennan (eds.) *Sea-Level Changes*. Basil Blackwell, Oxford, pp. 371–390.

Shennan, I. and Woodworth, P.L. (1992) A comparison of late Holocene and twentieth-century sea-level trends from the UK and North Sea region. *Geophysical Journal International*, **109**, 96–105.

Shennan, I., Tooley, M.J., Davis, M.J. and Haggart, B.A. (1983) Analysis and interpretation of Holocene sea-level data. *Nature*, **302**, 404–406.

Shepard, F.P. (1963) Thirty-five thousand years of sea level. In: *Essays in Marine Geology in Honor of K.O. Emery*. University of South California Press, Los Angeles, pp. 1–10.

Sirocko, F., Sarnthein, M., Erlenkeuser, H., Lange, H., Arnold, M. and Duplessy, J.C. (1993) Century-scale events in monsoonal climate over the past 24,000 years. *Nature*, **364**, 322–324.

Smith, A.J. (1989) Archaeology and sea-level change in the southwestern Pacific: no simple story. In: D.B. Scott, P.A. Pirazzoli and C.A. Honig (eds.) *Late Quaternary Sea-Level Correlation and Applications*. Kluwer, Dordrecht, NATO ASI Series C, vol. 256, pp. 195–206.

Smith, D.E., Cullingford, R.A. and Haggart, B.A. (1985) A major coastal flood during the Holocene in eastern Scotland. *Eiszeitalter und Gegenwart*, **35**, 109–118.

Solheim, A., Riis, F., Elverøi, A., Faleide, J.I., Jensen, L.N. and Cloetingh, S. (eds.) (1996) *Impact of Glaciations on Basin Evolution: Data and Models from the Norwegian Margin and Adjacent Areas*. Global and Planetary Change, **12** (1–4), 1–448.

Sowers, T. and Bender, M. (1995) Climate records covering the last deglaciation. *Science*, **269**, 210–214.

Stanley, D.J. and Warne, A.G. (1994) Worldwide initiation of Holocene marine deltas by deceleration of sea-level rise. *Science*, **265**, 228–231.

Stea, R.R. (ed.) (1987) *Quaternary Glaciations, Geomorphology and Sea-Level Changes: Bay of Fundy Area*. NATO ASI and IGCP Project 200 Symposium Field Trip, July 20–26. Dalhousie University, Halifax, 79 pp.

Stephenson, T.A. and Stephenson, A. (1949) The universal features of zonation between tide-marks on rocky coasts. *Journal of Ecology*, **37** (2), 289–305.

Stephenson, T.A. and Stephenson, A. (1972) *Life Between Tidemarks on Rocky Shores*. W.H. Freeman, San Francisco, 425 pp.

Stewart, I.S. and Hancock, P.L. (1993) Neotectonics. In: Hancock, P.L. (ed.) *Continental Deformation*. Pergamon Press, Oxford, pp. 370–409.

Stewart, R.W. (1989). Sea-level rise or coastal subsidence? *Atmosphere-Ocean*, **27** (3), 461–477.

Stiros, S.C., Arnold, M., Pirazzoli, P.A., Laborel, J., Laborel, F. and Papageorgiou,

S. (1992) Historical coseismic uplift on Euboea Island, Greece. *Earth and Planetary Science Letters*, **108**, 109–117.

Streif, H. (1979–1980) Cyclic formation of coastal deposits and their indications of vertical sea-level changes. In Groupe NIVMER (ed.) *Les Indicateurs de Niveaux Marins*. Oceanis, vol. 5, Fasc. Hors-Série, pp. 303–306.

Stuiver, M. and Braziunas, T.F. (1993) Modeling atmospheric ^{14}C influences and ^{14}C ages of marine samples to 10,000 BC. *Radiocarbon*, **35**, 137–189.

Stuiver, M. and Reimer, P.J. (1986) A computer program for radiocarbon age calibration. *Radiocarbon*, **28** (2B), 1022–1030.

Suess, E. (1885) *Das Antlitz der Erde*. Wien.

Suter, J.R. (1986) Buried late Quaternary fluvial channels on the Louisiana continental shelf, USA. *Journal of Coastal Research, Special Issue*, **1**, 27–37.

Tavernier, R. and Moormann, F. (1954) Les changements du niveau de la mer dans la plaine maritime flamande pendant l'Holocène. *Geologie en Mijnbouw*, **16**, 201–206.

Taylor, F.W., Isacks, B.L., Jouannic, C., Bloom, A.L. and Dubois, J. (1980) Coseismic and Quaternary vertical tectonic movements, Santo and Malekula Islands, New Hebrides Island arc. *Journal of Geophysical Research*, **85**, 5367–5381.

Thiede, J. (1978) A glacial Mediterranean. *Nature*, **276**, 680–683.

Thommeret, Y., Thommeret, J., Laborel, J., Montaggioni, L.F. and Pirazzoli, P.A. (1981) Late Holocene shoreline changes and seismo-tectonic displacements in western Crete (Greece). *Zeitschrift für Geomorphologie, Supplement*, **40**, 127–149.

Thompson, L.G., Mosley-Thompson, E., Davis, M.E., Lin, P.N., Henderson, K.A., Cole-Dai, J., Bolzan, J.F. and Liu, K.B. (1995) Late glacial stage and Holocene tropical ice core records from Huascarán, Peru. *Science*, **269**, 46–50.

Thomsen, H. (1982) Late Weichselian shore-level displacement on Nord-Jæren, south-west Norway. *Geologiska Föreningens i Stockholm Förhandlingar*, **103**, 447–468.

Thomson, R.E. and Tabata, S. (1987) Steric height trends at ocean station PAPA in the northeast Pacific Ocean. *Marine Geodesy*, **2**, 103–113.

Tooley, M.J. (1982) Sea-level changes in northern England. *Proceedings of the Geologists' Association*, **93**, 43–51.

Torunski, H. (1979) Biological erosion and its significance for the morphogenesis of limestone coasts and for nearshore sedimentation (Northern Adriatic). *Senckenbergiana Maritima*, **11**, 193–265.

Toyoshima, Y. (1965) Some wave-cut features in granitic regions, Uradome coast, Tottori Prefecture. *Tottori daigaku gakugeigakubu Kenkyû hôkoku (Liberal Arts Journal, Tottori University)*, **16**, 46–59 (in Japanese with English abstract).

Trudgill, S.T. (1976) The marine erosion of limestones on Aldabra Atoll, Indian Ocean. *Zeitscrift für Geomorphologie, Supplement*, **26**, 164–200.

Trupin, A. and Wahr, J. (1990) Spectroscopic analysis of global tide gauge sea level data. *Geophysical Journal International*, **100**, 441–453.

UNESCO (1971) Scientific framework of world water balance. *Technical Papers Hydrology*, **7**, 27 pp.

United States Department of Energy (1985) *Glaciers, Ice Sheets, and Sea Level: Effect of CO$_2$-induced Climatic Change*. National Technical Information Service, United States Department of Commerce, Springfield, VA, Report DOE/ER/60235-1, 348 pp.

Valentin, H. (1952) Die Küsten der Erde. *Petermanns Geographische Mitteilungen, Ergänzungsband*, **246**, 118 pp.

Van Andel, T.H. and Veevers, J.J. (1967) Morphology and sediments of the Timor

Sea. *Bulletin of the Bureau of Mineral Resources, Geology and Geophysics, Australia*, **83**, 1–173.

Van Campo, E., Duplessy, J.C., Prell, W.L., Barratt, N. and Sabatier, R. (1990) Comparison of terrestrial and marine temperature estimates for the past 135 kyr off southeast Africa: a test for GCM simulations of palaeoclimate. *Nature*, **348**, 209–212.

Van de Plassche, O. (1982) Sea-level change and water-level movements in the Netherlands during the Holocene. *Mededelingen Rijks Geologiske Dienst*, **36** (1), 93 pp.

Van de Plassche, O. (1991) Late Holocene sea level fluctuations on the shore of Connecticut inferred from transgressive and regressive overlap boundaries in salt-marsh deposits. *Journal of Coastal Research, Special Issue*, **11**, 159–179.

Van de Plassche, O. and Roep, T.B. (1989) Sea-level changes in the Netherlands during the last 6500 years: basal peat vs. coastal barrier data. In: D.B. Scott, P.A. Pirazzoli and C.A. Honig (eds.) *Late Quaternary Sea-Level Correlation and Applications*. Kluwer, Dordrecht, NATO ASI Series C, vol. 256, pp. 41–56.

Van Veen, J. (1954) Tide-gauges, subsidence-gauges and flood-stones in the Netherlands. *Geologie en Mijnbouw*, **16**, 214–219.

Veeh, H.H. and Veevers, J.J. (1970) Sea level at −175 m off the Great Barrier Reef 13,600 to 17,000 year ago. *Nature*, **226**, 536–537.

Vita-Finzi, C. (1986) *Recent Earth Movements*. Academic Press, London, 226 pp.

Walcott, R.I. (1970a) Flexural rigidity, thickness, and viscosity of the lithosphere. *Journal of Geophysical Research*, **75** (20), 3941–3954.

Walcott, R.I. (1970b) Flexure of the lithosphere at Hawaii. *Tectonophysics*, **9**, 435–446.

Walcott, R.I. (1972) Past sea levels, eustasy and deformation of the earth. *Quaternary Research*, **2**, 1–14.

Warrick, R. and Oerlemans, J. (1990) Sea level rise. In: J.T. Houghton, G.J. Jenkins and J.J. Ephraums (eds.) *Climate Change – The IPCC Scientific Assessment*. Cambridge University Press, Cambridge, pp. 257–281.

Watson, R.T., Rodhe, H., Oeschger, H. and Siegenthaler, U. (1990) Greenhouse gases and aerosols. In J.T. Houghton, G.J. Jenkins and J.J. Ephraums (eds.) *Climate Change – The IPCC Scientific Assessment*. Cambridge University Press, Cambridge, pp. 1–40.

Webb, T. III, Kutzbach, J. and Street-Perrott, F.A. (1985) 20,000 years of global climatic change: paleoclimatic research plan. In: T.F. Malone and J.G. Roederer (eds.) *Global Change*. ICSU Press Symposium Series 5, Cambridge University Press, Cambridge, pp. 182–218.

Wellman, H.W. (1967) Tilted marine beach ridges at Cape Turakirae, New Zealand. *Journal of Geosciences, Osaka City University*, **10**, 123–129.

White, W., Hasunuma, K. and Meyers, G. (1979) Large-scale secular trend in steric sea level over the western North Pacific from 1954–1974. *Journal of the Geodetic Society of Japan*, **25** (1), 49–55.

Wiggers, A.J. (1954) Compaction of sediments older than Holocene in relation to the subsidence of the Netherlands. *Geologie en Mijnbouw*, **16**, 179–184.

Wigley, T.M.L. and Raper, S.C.B. (1987) Thermal expansion of sea water associated with global warming. *Nature*, **330**, 127–131.

Wigley, T.M.L. and Raper, S.C.B. (1992) Implications for climate and sea levels of revised IPCC emissions scenarios. *Nature*, **357**, 293–300.

Wingfield, R.T.R. (1995) Major themes identified at a meeting on "Continental shelf evidence of sealevels over the last 20 ka". *Quaternary Research Association Annual Meeting*, Edinburgh, 2 pp.

Woodroffe, C. (1988) Mangroves and sedimentation in reef environments: indicators of past sea-level changes, and present sea-level trends? *Proceedings of Sixth International Coral Reef Symposium, Australia*, vol. 3, pp. 535–539.

Woodroffe, C. (1990) The impact of sea-level rise on mangrove shorelines. *Progress in Physical Geography*, **14** (4), 483–520.

Woodroffe, C. (1992) Mangrove sediments and geomorphology. In: A.I. Robertson and D.M. Alongi (eds.) *Tropical Mangrove Ecosystems*. Coastal and Estuarine Studies vol. 41, American Geophysical Union, Washington DC, pp. 7–41.

Woodroffe, C. and Grindrod, J. (1991) Mangrove biogeography: the role of Quaternary environmental and sea-level change. *Journal of Biogeography*, **18**, 479–492.

Woodroffe, C.D., Thom, B.G., Chappell, J. and Head, J. (1987) Relative sea levels in the South Alligator river region, North Australia, during the Holocene. *Search*, **18** (4), 198–200.

Woodworth, P.L. (1990) A search for accelerations in records of European mean sea level. *International Journal of Climatology*, **10**, 129–143.

Yang, H. and Xie, Z. (1984) Sea-level changes in East China over the past 20,000 years. In: R.O. Whyte (ed.) *The Evolution of Eastern Asian Environment*. University of Hong Kong, pp. 288–308.

Yonekura, N. (1972) A review on seismic crustal deformations in and near Japan. *Bulletin of the Department of Geography, University of Tokyo*, **4**, 17–50.

Yonekura, N. (1975) Quaternary tectonic movements in the outer arc of southwest Japan with special reference to seismic crustal deformations. *Bulletin of the Department of Geography, University of Tokyo*, **7**, 19–71.

Yoshikawa, T., Kaizuka, S. and Ota, Y. (1981) *The Landforms of Japan*. University of Tokyo Press, 222 pp.

Zendrini, A. (1802) Sull'alzamento del livello del mare. *Giornale dell'Italiana Letteratura (Padova)*, **2**, 3–37.

Zerbini, S. et al. (16 authors) (1996) Sea level in the Mediterranean: a first step towards separating crustal movements and absolute sea-level variations. *Global and Planetary Change*, **14** (1–2), 1–48.

Zhao, S. and Zhao, X. (1986) The progress of study in sea-level changes of the late Quaternary, China. In: *China Sea Level Changes*. China Ocean Press, Beijing, pp. 28–34.

Zhao, X., Geng, X. and Zhang, J. (1982) Sea level changes in Eastern China during the past 20,000 years. *Acta Oceanologica Sinica*, **1**, 248–258.

Author Index

Geographical Index

Subject Index